Translational Systems Sciences

Volume 26

Kyoichi Kijima • Junichi Iijima • Ryo Sato •
Hiroshi Deguchi • Bumpei Nakano
Editors

Systems Research I

Essays in Honor of Yasuhiko Takahara on Systems Theory and Modeling

Springer

Editors

Kyoichi Kijima
School of Business Management
Bandung Institute of Technology
Kawasaki, Japan

Junichi Iijima
School of Management
Tokyo University of Science
Tokyo, Japan

Ryo Sato
School of Management
Tokyo University of Science
Tokyo, Japan

Hiroshi Deguchi
Faculty of Commerce and Economics
Chiba University of Commerce
Tokyo, Japan

Bumpei Nakano
Yokohama, Japan

ISSN 2197-8832 ISSN 2197-8840 (electronic)
Translational Systems Sciences
ISBN 978-981-16-9936-8 ISBN 978-981-16-9937-5 (eBook)
https://doi.org/10.1007/978-981-16-9937-5

This Springer imprint is published by the registered company Springer Nature Singapore Pte Ltd.
The registered company address is: 152 Beach Road, #21-01/04 Gateway East, Singapore 189721, Singapore

Preface

It is a great pleasure to publish Systems Research I and II: Essays in Honor of Yasuhiko Takahara, to commemorate the 50th anniversary of Dr. Yasuhiko Takahara's research and education activities, who has been active at the world level in the field of systems research. We compile representative research of researchers and practitioners scented by Dr. Takahara from Japan and abroad who pay homage and gratitude to him into two volumes.

Dr. Takahara completed his research with Dr. M. Mesarovic at Case Western Reserve University. He brought up the results of General Systems Theory (GST), especially Mathematical General Systems Theory (MGST), from the United States to Japan to be appointed to Tokyo Institute of Technology in 1972. In Japan, it was a time the term "system engineering" began to attract interest in gradually establishing the system as a unique field and capturing the essence of what is recognized as a system.

We, editors-in-chief, who fortunately shared that era, still remember the shock we had when we first learned GST, especially MGST. MGST attempts to transparently understand the properties of systems such as interdependency and emergent property, by discussing the logic related to systems in a set-theoretic framework to formulate causal systems and hierarchical systems.

A wide range of books used as textbooks at the seminars in Takahara laboratory remind us of such good old days. They not only have worked as the soils supporting our research since then but also certainly reflect some part of the background of knowledge at that time. They include:*Universal Algebra* (George Graetzer); *Algebra* (Saunders MacLane and Garrett Birkoff); *Introduction to Topology and Modern Analysis* (George F. Simmons); *Topology* (James Dugundji); *Abstract and Concrete Categories: The Joy of Cats* (Jiri Adámek, Horst Herrlich, and George E. Strecker); *Model Theory* (C.C. Chang, H. Jerome Keisler), *Beginning Model Theory* (Jane Bridge); *Model-Based Systems Engineering* (A. Wayne Wymore); *Theories of Abstract Automata* (Michael A. Arbib); *The Specification of Complex Systems* (B. Cohen et al.); *The Structure of Scientific Theories* (Frederick Suppe (Editor)), *Goedel, Escher, Bach* (Douglas R. Hofstadter); *Forever Undecided* (Raymond Smullyan); *Introduction to Systems Philosophy* (Ervin Laszlo), *Living Systems*

(James Grier Miller); *Facets of Systems Science* (George J. Klir), *Systems Thinking Systems Practice* (Peter Checkland); *Heuristics* (Judea Pearl); *Multifaceted Modelling and Discrete Event Simulation* (Bernard P. Zeigler).

Over the 50 years since then, Dr. Takahara has developed the concept and theory of general systems centered on formal system research. By projecting mathematical general systems theory from meta-theory to the real world, he has powerfully promoted diverse but coherent systems research with a strong desire and intention to construct knowledge not only in theory (episteme) but also in engineering (techne) and practice (phronesis). They include systems modeling, information systems, decision support systems, and systems thinking. The results and findings there are now having a great impact on improving our socio-economic situations, which are becoming more and more complex, by providing the basis of ideas for the sophistication of business models and creation of new services by networks and ICT.

At the same time, Dr. Takahara has promoted educational activities with these studies and research as a nursery to give a great influence on many researchers and practitioners, not limited to students who received his direct scent. It triggers intellectual excitement and acts as a device to encourage their diversified but coherent systems study.

The Takahara School of GST is outstanding in its interdisciplinary and transdisciplinary approach. As you can see in the table of contents of the book, it ranges from highly abstract mathematics to practical applications of social science. It is very fortunate in systems research history that the Takahara school has created such a wide range of content in a deep dialogue using "system" as the common concept.

Some 20 authors gathered here have established their position by developing systems concepts in theory, models, methodologies, and applications in various ways, keeping in their mind the works by Dr. Takahara and co-authored with Mesarovic. Their activities include:

- Research to develop the strong intellectual desire for generality tackled by mathematical general systems theory into knowledge to connect different levels and approaches to the same object for solving actual problems. They are eager to catch the spirit that GST aimed at in the early days.
- Enterprise by which MGST and system engineering are practically fused by examining the basic concept of the system to develop knowledge applicable to hot topics these days.
- Development of what we call translational approach that connects theory and practice in a cyclic way, which follows the process of analyzing mechanisms, solving in an evidence-based way, and intervening in a problematic situation.
- Exploration of new disciplines such as Decision Systems Science and Service Systems Science by accommodating "hard" systems theory with "soft" systems thinking.
- Promotion of a formal approach in the field of Information Systems, for example, to identify isomorphisms between different recommendation systems and decision schemes known in social choice theory, and to the Enterprise Ontology in Design and Engineering Methodology for Organization.

Since the articles contributed by the twenty authors have a wide range of issues, we decided to structure them based on topics and approaches and publish them as a two-volume set. The first volume consists of 11 chapters divided into two parts dealing with the field of Systems Theory and Modeling.

Finally, we would like to express deep thanks to you Yutaka Hirachi of Springer Japan and Selvakumar Rajendran, and other staff of SPi Global. Without their patience and understanding, this volume could not be published.

Kawasaki, Japan — Kyoichi Kijima
Tokyo, Japan — Junichi Iijima
Tokyo, Japan — Ryo Sato
Tokyo, Japan — Hiroshi Deguchi
Yokohama, Japan — Bumpei Nakano
December 2021

Contents

Part I
General Systems Theory

Chapter 1
Mesarovic-Takahara Time Systems Under the Effect of Feedback Mechanism

Jeffrey Yi-Lin Forrest, Zhen Li, Yohannes Haile, and Liang Xu

Abstract This chapter focuses on the study of Mesarovic-Takahara (MT) time systems and demonstrates how various properties of such systems are feedback invariant. In particular, after a brief introduction of the concepts of MT time systems and feedback systems, this work shows how these concepts play an important role in establishing the relationship between manufacturing and further industrial transformations. With such practical importance of feedback systems established, this presentation turns its focus to the theoretical study of various feedback invariant properties of MT systems. As an application, the concept of feedback systems is employed to explore when and how government economic policies can become effective in terms of assisting with and stimulating economic growth.

Keywords Attractor · Chaos · Consumer surplus · Economic system · Government policy · Market demand · Stationary system · Time-invariable realization

---

J. Y.-L. Forrest (✉)
Department of Accounting Economics Finance, Slippery Rock University, Slippery Rock, PA, USA
e-mail: Jeffrey.forrest@sru.edu

Z. Li
College of Business, Texas Woman's University, Denton, TX, USA
e-mail: zli@twu.edu

Y. Haile
Black School of Business and School of Engineering, Pennsylvania State University Behrend, Erie, PA, USA
e-mail: yxh313@psu.edu

L. Xu
Department of Management and Marketing, Slippery Rock University, Slippery Rock, PA, USA
e-mail: liang.xu@sru.edu

K. Kijima et al. (eds.), *Systems Research I*, Translational Systems Sciences 26,
https://doi.org/10.1007/978-981-16-9937-5_1

1.1 Introduction

At this special moment of honoring one of the great thinkers of our time, the first coauthor of this chapter likes to express his sincere and heartfelt appreciation to Professor Yasuhiko Takahara. It is Professor Takahara's publication of various important monographs and papers that majorly shaped this coauthor's entire professional career since the mid-1980s. As a tribute to Professor Takahara, this chapter continues the scholarly works started by him in the field of systems research.

To achieve this research objective, this chapter investigates various classes of Mesarovic-Takahara (MT) time systems, such as those that are strongly stationary or precausal or causal, and various important properties, such as time-invariable realization, chaos, and attractors. The focus is on the effect of the feedback mechanism on these particular systems and properties. Other than continuing the tradition of Professor Takahara's works in the language of set theory, this chapter employs recent developments in the research of economics to illustrate how some of the derived set-theoretic results can be beneficially applied to enrich our knowledge on

- How races between market exchange and manufacturing can be purposefully employed for a nation to kick-start and maintain a self-sustaining momentum of growth.
- Under what conditions government economic policies can be effective in terms of economic development.

The rest of this chapter is organized as follows: Sect. 1.2 introduces the basic concepts needed for the work and demonstrates how feedback systems function in real life. Section 1.3 features the main properties of MT time systems, such as strong stationality, (pre-)causality, time-invariant realization, chaos, and attractors. Following this theoretical study, Section 1.4 demonstrates how some of the basic concepts and established conclusions of MT time systems can be practically employed to study when government economic policies become effective. Section 1.5 concludes this presentation.

1.2 Basic Concepts

This section presents the basic terms needed for the rest of this research. After introducing the fundamental terminologies in the first subsection, such as the concepts of input-output systems and MT time systems, the second subsection illustrates the practical importance of feedback systems by examining how such a systemic structure appears naturally within a growing economy.

1.2.1 The Underlying Concepts

Let X and Y be two sets. Then the following binary relation S

$$\varnothing \neq S \subset X \times Y$$

is known as an input-output system (Mesarovic & Takahara, 1975) with X being the input space and Y the output space of S, respectively. It is evident that most economic entities in the business world are input-output systems, because various inputs are needed for these entities to survive and particular outcomes are offered to the marketplace (Forrest, 2010; Porter, 1985).

Let the positive half of the real number line, written as $T = [0, +\infty)$, be the time axis, and A and B be two linear spaces over the same field $\mathcal{A}$. Define sets A^T and B^T as follows:

$$A^T = \{f : f \text{ is a mapping } T \to A\} \quad \text{and} \quad B^T = \{f : f \text{ is a mapping } T \to B\}.$$

As conventionally known, these sets A^T and B^T become linear spaces over $\mathcal{A}$ if the following operations are introduced. For any elements $f, g \in A^T$ (respectively, $\in B^T$), and any scalar$\alpha \in \mathcal{A}$,

$$(f + g)(t) = f(t) + g(t) \quad \text{and} \quad (\alpha f)(t) = \alpha \cdot f(t), \quad \text{for each } t \in T.$$

Each input-output system S defined on A^T and B^T, or symbolically, $S \subset A^T \times B^T$, is referred to as an Mesarovic-Takahara (MT) time system (Mesarovic & Takahara, 1989).

Let $\mathcal{A}$ be a field, A and B linear spaces over $\mathcal{A}$, and S an input-output system satisfying that.

1. $\varnothing \neq S \subset A \times B$
2. $s \in S$ and $s' \in S$ imply $s + s' \in S$
3. $s \in S$ and $\alpha \in \mathcal{A}$ imply $\alpha \bullet s \in S$

where $+$ and $\cdot$ are addition and scalar multiplication in $A \times B$, respectively, defined as follows: For any $(x_1, y_1), (x_2, y_2) \in A \times B$ and any $\alpha \in \mathcal{A}$,

$$(x_1, y_1) + (x_2, y_2) = (x_1 + x_2, y_1 + y_2) \quad \text{and} \quad \alpha (x_1, y_1) = \left(\alpha x_1, \alpha y_1\right),$$

then S is then known as a linear system (Forrest, 2010).

Let $S \subset A \times B$ be a linear system and $S_f : B \to A$ a linear function or known as a functional system. Then the feedback system S' of S by S_f is defined (Forrest, 2010) as follows: for any $(x, y) \in A \times B$

$$(x, y) \in S' \leftrightarrow (\exists z \in A)\left((x + z, y) \in S \quad \text{and} \quad (y, z) \in S_f\right). \tag{1.1}$$

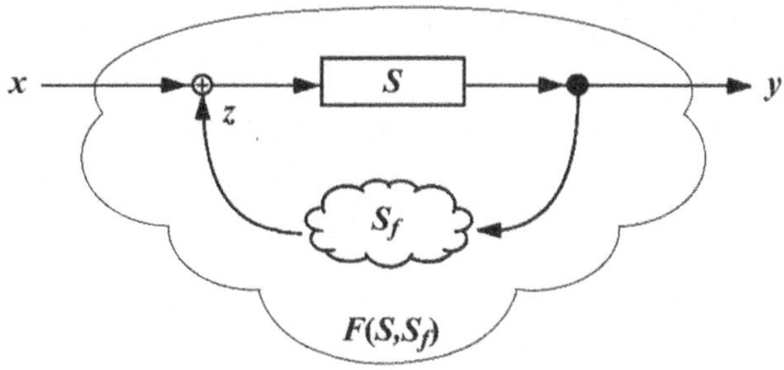

Fig. 1.1 The structure of a feedback system

System S is known as an original system; and S_f as a feedback component system. Let

$$\mathcal{S} = \{S \subset A \times B : S \text{ is a linear system}\},$$

$$\mathcal{S}_f = \{S_f : B \to A : S_f \text{ is a linear functional system}\},$$

and

$$\mathcal{S}' = \left\{S' \subset A \times B : S' \text{ is a subset}\right\}.$$

Then a feedback transformation $F : \mathcal{S} \times \mathcal{S}_f \to \mathcal{S}'$ is defined by Eq. (1.1), and known as the feedback transformation over the linear spaces A and B (Forrest, 2010; Lin & Ma, 1990; Saito, 1986). Figure 1.1 shows the geometric meaning of the concept of feedback systems, where the output y is jointly affected by the input x and the feedback loop $F(S, S_f)$.

Scholars from different disciplines have employed the concept of feedback to develop important insights and conclusions (Bayliss, 1966; Deng, 1985; Forrest et al., 2018a, b; Henig, 1983; Milsum, 1966; Negoita, 1992; Saito, 1986, 1987; Saito & Mesarovic, 1985; Takahara & Asahi, 1985; Wonham, 1979; Wu, 1981; Zadeh, 1965). The definition of general feedback systems is first introduced by Saito and Mesarovic (1985). For relevant theoretical studies, see Lin and Ma (1990).

1.2.2 How the Feedback Mechanism Functions in an Economy: An Example

To demonstrate the practical importance of the concept of feedback systems, this subsection shows how the feedback mechanism, as introduced above, organically associates market exchange and manufacturing for a nation that desires to kick-start and maintain its self-sustaining momentum of growth. By self-sustained momentum of growth, it means a self-sustained societal development that penetrates all economic, social, and political aspects of the society. As a result of such societal development, advanced technologies appear and alter how people live and how government operates. Meanwhile, a population explosion follows the societal development, while the methods of production and service are revolutionized (Heaton, 2017; Wen, 2016).

Beyond confirming what *consumers really want in life and satisfying market demands, market exchanges dictate the level of competitive supplies and necessitates needed innovations through demonstrating consumer demands* (Forrest et al., 2017). *The level of competitive supplies and the quality of innovations are determined by sufficient market depth (or purchasing power). They in turn directly further* the development of technologies due to intensified competitions. On the other hand, f*or a nation to develop its needed market depth, it mobilizes a portion of the available labor force, freed from producing life necessities, into the manufacturing of luxurious goods.* As a result of the relocation of labor, citizens have much more incomes than before, which raises their purchasing powers. That helps deepen the product market. That is, market competition encourages the development of new technologies and helps industries advance to higher levels. At the same time, these developments simultaneously expand various markets, be they financial, product, or labor, with greater varieties of products. For more discussions along this line, see Forrest (2010).

This discussion naturally reveals how a feedback mechanism plays its role in a nation's economic development. In particular, let us visualize the input x as a particular government policy and y the corresponding economic output produced by adopting the policy. It is intuitive to see that although the output y is a consequence of the policy x, it represents the combined effect of the policy and the market reaction, denoted by $S_f(S(x))$, to the policy. In reality, of course, the situation can take one of two possible scenarios—the economy is a relatively closed system from the outside world or an open system. For the former case, this discussion does not involve any externalities and related costs, while for the latter case, the market reaction $S_f(S(x))$ includes all those from domestic and foreign markets and the environment.

Speaking differently, the previous paragraph indicates that the race between market exchange and manufacturing production is indeed a feedback system, within which market demands excite manufacturers to produce more and better products and entrepreneurs to introduce new and novel offers. Such a feedback mechanism forces firms to satisfy the increasing need of production by hiring additional employees with rising salaries. The race simultaneously reinforces the market depth and the demand of the market. Such mechanism underlying the working of the described feedback system is figuratively shown in Fig. 1.2.

One good example that illustrates the abstract discussion above is the phenomena of industrial revolutions that occurred one after another in the recent history from around the world (Forrest et al. 2018a, b; Rostow, 1960; Wen, 2016). In each of these occasions, it was the intensifying race between market exchanges and manufacturing productions that brought forward with the desired self-sustaining

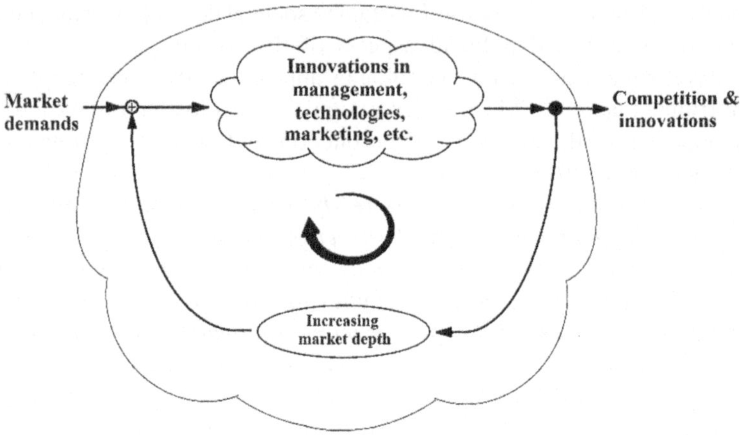

Fig. 1.2 The feedback race between market exchange and manufacturing

momentum of economic growth. However, contrary to this example of positive confirmation, the third industrial revolution in the United States did not really help the US economy enjoy rising salaries even though the economy nearly reached its full employment in February 2020. The reason behind the appearance of this counter-intuitive situation is that within the ongoing economic globalization, the US economy represents merely a local one, while what is discussed above expresses what holds true within the entire economic system that entertains an intensifying race between market exchanges and manufacturing productions. More specifically, instead of the US economy, it is those foreign economies involved in relevant cross-national productions that enjoy the fruitful benefits of the third industrial revolution that originated in the USA.

1.3 Properties of MT Time Systems

This section presents results on how the feedback transformation affects various properties of linear time systems, such as strong stationality, (pre-)causality, time-invariable realization, chaos, and attractors.

1.3.1 Linear Time Systems That Are Strongly Stationary

In this section, all symbols and their relevant assumed conditions are maintained as given previously. Then, the following result implies that the feedback transformation

$F : \mathcal{S} \times \mathcal{S}_f \rightarrow \mathcal{S}'$, given in Eq. (1.1), is well defined on the class of all MT time systems.

Theorem 1.1 *If $S \subset A \times B$ is a linear system and $S_f : B \rightarrow A$ a linear functional system, then the feedback system $F(S, S_f) \subset A \times B$ is also a linear system.*

Proof For arbitrarily chosen $(x_1, y_1), (x_2, y_2) \in F(S, S_f)$ and $\alpha \in \mathcal{A}$, Eq. (1.1) implies that there are $z_1, z_2 \in A$ satisfying

$$l\,(x_1 + z_1, y_1) \in S \quad \text{and} \quad (y_1, z_1) \in S_f \quad \text{and} \quad (x_2 + z_2, y_2) \in S \quad \text{and} \quad (y_2, z_2) \in S_f. \tag{1.2}$$

Because both S and S_f are linear systems, Eq. (1.2) implies

$$\left(\alpha x_1 + \alpha z_1, \alpha y_1\right) \in S \quad \text{and} \quad \left(\alpha y_1, \alpha z_1\right) \in S_f,$$

from which we have $\alpha(x_1, y_1) = (\alpha x_1, \alpha y_1) \in F(S, S_f)$.

Additionally, because of the linearity of S and S_f, Eq. (1.2) implies

$$((x_1 + x_2) + (z_1 + z_2), (y_1 + y_2)) \in S \quad \text{and} \quad ((y_1 + y_2), (z_1 + z_2)) \in S_f.$$

Hence, we have $(x_1, y_1) + (x_2, y_2) = (x_1 + x_2, y_1 + y_2) \in F(S, S_f)$.

By combining what are established above, we conclude that the feedback system $F(S, S_f)$ is a linear system over $A \times B$. QED

For two arbitrary time moments $t, t' \in T$, satisfying $t < t'$, we denote the following intervals abbreviately

$$T^t = \left[0, t\right), \quad T_t = [t, +\infty), \quad T_{tt'} = \left[t, t'\right), \quad \overline{T}^t = [0, t], \quad \overline{T}_{tt'} = \left[t, t'\right].$$

Accordingly, the restrictions of any $x \in A^T$ with respect to these intervals of time are respectively denoted by

$$x^t = x\left|T^t\right., \quad x_t = x\,|T_t\,, \quad x_{tt'} = x\,|T_{tt'}\,, \quad \overline{x}^t = x\left|\overline{T}^t\right., \quad \overline{x}_{tt'} = x\left|\overline{T}_{tt'}\right..$$

More generally, for sets and vectors similar notations will be utilized. For instance, for $(x, y) \in A^T \times B^T$, we denote

$$(x, y)^t = \left(x^t, y^t\right), \quad (x, y)_t = (x_t, y_t), \quad \text{etc.}$$

A subset $S \subset X \times Y \subset A^T \times B^T$ is called a linear time system (Lin, 1989; Lin & Ma, 1987; Ma & Lin, 1992) if S is a linear subspace of $X \times Y$. Let the domain of S be written as $D(S)$, known as the input space of S. It is defined as follows:

$$D(S) = \{x \in X : \exists y \in Y \ni (x, y) \in S\}.$$

Without loss of generality, assume that $D(S) = X$ and $R(S) = Y$, where $R(S)$ is the range of S, known as the output space of S, defined by

$$R(S) = \{y \in Y : \exists x \in X \ni (x, y) \in S\}.$$

For any $x, x' \in X$ and any $t \in T$, assume that $x^t \circ x'_t \in D(S)$, where $x^t \circ x'_t$ is the concatenation of x^t and x'_t, which is defined as follows: for $s \in T$,

$$x^t \circ x'_t(s) = \begin{cases} x(s), & \text{if } s < t \\ x'(s), & \text{if } s \geq t \end{cases}.$$

For a given real number τ, let σ^τ be the shift operator defined on X as follows: for any $x \in X$, $\sigma^\tau(x) \in X$ satisfying that

$$\sigma^\tau(x)(\xi) = x(\xi - \tau), \quad \forall \xi \in T_\tau.$$

Figure 1.3a–c show the geometric meaning of the concept of the shift operator σ^τ respectively for the cases when $\tau = 0$, $\tau > 0$, and $\tau < 0$. In Fig. 1.3b, c the dotted $A \times T$ planes indicate the locations of the $A \times T$ plane before the shift operation σ^τ is applied. In other words, if $\tau > 0$, the shift operator σ^τ moves the graph of time function x τ units to the right; if $\tau < 0$, the shift operator σ^τ emphasizes on the portion of the graph of x on the right of the vertical line $t = \tau$

When a linear time system $S \subset X \times Y$ satisfies

$$\forall t \in T \left(\sigma^{-t}\left(S \mid T_t\right) = S\right),$$

then S is said to be strongly stationary (Lin, 1989; Lin & Ma, 1987; Ma & Lin, 1992).

Theorem 1.2 *Assume that $S \subset X \times Y$ is a linear time system and $S_f : Y \to X$ a strongly stationary linear functional time system. Then a sufficient and necessary condition for the feedback system $F(S, S_f)$ to be strongly stationary is that system S is strongly stationary.*

Proof (⇒). For the proof of this necessity, the feedback system $F(S, S_f)$ is assumed to be strongly stationary. Then the following holds true:

$$S = \{(x + S_f(y), y) : (x, y) \in F(S, S_f)\}.$$

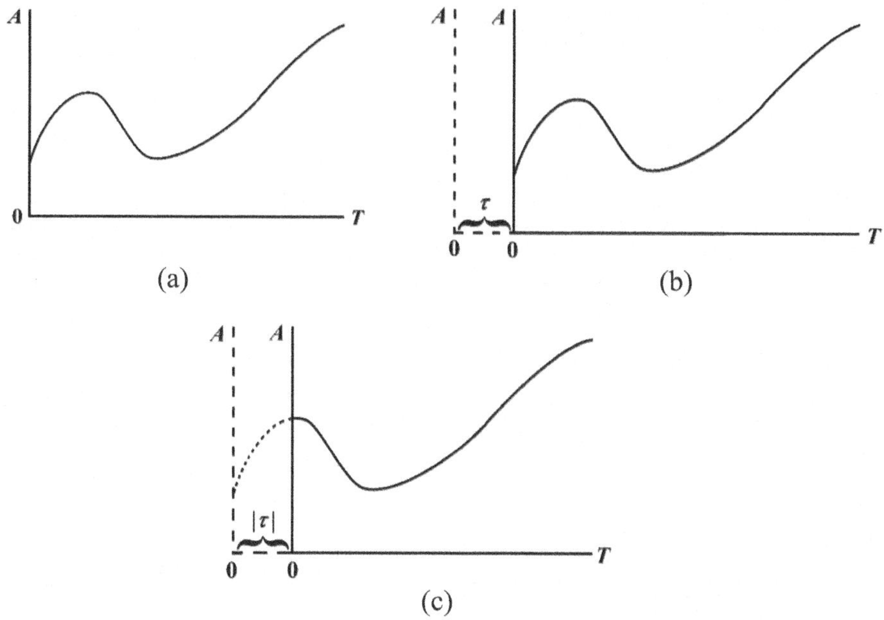

Fig. 1.3 How the shift operator σ^τ affects the graph of x. (**a**) The original x in plane $T \times A$. (**b**) After the shift σ^τ is applied if $\tau > 0$. (**c**) After the shift σ^τ is applied if $\tau < 0$

To complete the proof of this part, we only need to show that

$$\forall (x, y) \in S \forall t \in T \left(\sigma^{-t}\left((x, y) | T_t\right) \in S\right).$$

To demonstrate this end, let $(x, y) \in S$ and $t \in T$. Then, we have

$$\begin{aligned}\sigma^{-t}\left((x, y) | T_t\right) &= \sigma^{-t}\left((x - S_f(y) + S_f(y), y) | T_t\right) \\ &= \left(\sigma^{-t}\left((x - S_f(y)) | T_t\right) + \sigma^{-t}\left(S_f(y) | T_t\right), \sigma^{-t}\left(y | T_t\right)\right).\end{aligned}$$

Because the systems $F(S, S_f)$ and S_f are assumed to be strongly stationary, we have

$$\left(\sigma^{-t}\left((x - S_f(y)) | T_t\right), \sigma^{-t}\left(y | T_t\right)\right) \in F\left(S, S_f\right),$$

and

$$\left(\sigma^{-t}\left(y | T_t\right), \sigma^{-t}\left(S_f(y) | T_t\right)\right) \in S_f. \tag{1.3}$$

Let $x' = \sigma^{-t}((x - S_{\text{f}}(y))|T_t)$ and $y' = \sigma^{-t}(y|T_t)$. Then, Eq. (1.3) indicates that $S_{\text{f}}(y') = \sigma^{-t}(S_{\text{f}}(y)|T_t)$. That is, we have

$$\sigma^{-t}\left((x, y)|\, T_t\right) = \left(x' + S_{\text{f}}\left(y'\right), y'\right), \tag{1.4}$$

where $(x', y') \in F(S, S_{\text{f}})$. So, Eq. (1.4) implies that $\sigma^{-t}((x, y)|T_t) \in S$.

($\Leftarrow$). This part of the sufficiency proof follows from the fact that for a linear functional system $S : X \to Y$, $F(S, S_f)$ is injective, for each arbitrarily $S_f \in \mathcal{S}_{\text{f}}$, if, and only if S is injective (Theorem 10.3 of Forrest, 2010) and the necessity part of this proof. QED

1.3.2 Linear Time Systems That Are Precausal or Causal

If a linear time system $S \subset X \times Y$ satisfies

$$(\forall t \in T)\,(\forall x \in X)\left(\overline{x}^t = \overline{0}^t \to R(S)\left|\overline{T}^t = S(0)\right|\overline{T}^t\right),$$

then S is referred to as a precausal system (Lin, 1989; Lin & Ma, 1987; Ma & Lin, 1992), where for $x \in X$, $S(x) = \{y \in Y : (x. y) \in S\}$.

Theorem 1.3 *A sufficient and necessary condition for a linear time system* $S \subset X \times Y$ *to be precausal is that* $(\forall t \in T)\,(\forall x, y \in X)\left(\overline{x}^t = \overline{y}^t \to S(x)\left|\overline{T}^t = S(y)\right|\left|\overline{T}^t\right.\right)$.

Proof ($\Rightarrow$). For this necessity part of the proof, S is assumed to be precausal. Hence, for any $x, y \in X$, we have

$$\overline{(x - y)}^t = \overline{0}^t \to S\,(x - y)\left|\overline{T}^t = S(0)\right|\left|\overline{T}^t\right..$$

This expression means that

$$\overline{x}^t = \overline{y}^t \to S(x)\left|\overline{T}^t = S(y)\right|\left|\overline{T}^t\right.,$$

because the condition $\overline{(x - y)}^t = \overline{0}^t$ is equivalent to that of $\overline{x}^t - \overline{y}^t = \overline{0}^t$. So, we have

$$\overline{x}^t = \overline{y}^t \text{ and } S\,(x - y)\left|\overline{T}^t = S(0)\right|\left|\overline{T}^t \to S(x)\right|\overline{T}^t - S(y)\left|\left|\overline{T}^t = S(0)\right|\right|\overline{T}^t.$$

Therefore, we have $S(x)\left|\overline{T}^t\right. = S(y)\left|\overline{T}^t\right. + S(0)\left|\overline{T}^t\right. = S(y)\left|\overline{T}^t\right.$.

($\Leftarrow$) For this part of the sufficiency proof, assumed is

$$(\forall t \in T)\,(\forall x, y \in X)\left(\overline{x}^t = \overline{y}^t \rightarrow S(x)\left|\overline{T}^t\right. = S(y)\left|\right|\overline{T}^t\right).$$

This assumption implies

$$(\forall t \in T)\,(\forall x \in X)\left(\overline{x}^t = \overline{0}^t \rightarrow S(x)\left|\overline{T}^t\right. = S(0)\left|\overline{T}^t\right.\right).\text{QED}$$

Theorem 1.4 *Assume that both $S : X \rightarrow Y$ and $S_{\mathrm{f}} : Y \rightarrow X$: Y are time systems that are linear functional causal. Then a sufficient and necessary condition for the feedback system $F(S, S_{\mathrm{f}})$ to be causal is that the time system $S \circ S_{\mathrm{f}} \circ F(S, S_{\mathrm{f}}) : D(F(S, S_{\mathrm{f}})) \rightarrow Y$ is causal.*

Proof The necessity condition follows from the fact that for any given causal functional time systems, their composition is also causal.

($\Leftarrow$). For the proof of the sufficiency condition, we assume that $S \circ S_{\mathrm{f}} \circ F(S, S_{\mathrm{f}})$ is a causal system. Hence, we have

$$\forall t \in T \forall x \in D\left(F\left(S, S_{\mathrm{f}}\right)\right)\left(\overline{x}^t = \overline{0}^t \rightarrow S \circ S_{\mathrm{f}} \circ F\left(S, S_{\mathrm{f}}\right)(x)\left|\overline{T}^t\right. = \overline{0}^t\right).$$

If we let $y = F(S, S_{\mathrm{f}})(x)$, we need to show that $\overline{y}^t = \overline{0}^t$. To this end, Eq. (1.1) implies

$$S\left(x + S_{\mathrm{f}}(y)\right) = y.$$

Therefore, we have $S(x) + S \circ S_{\mathrm{f}}(y) = y$, where

$$S(x)\left|\overline{T}^t\right. = \overline{0}^t \quad \text{and} \quad S \circ S_{\mathrm{f}}(y)\left|\overline{T}^t\right. = S \circ S_{\mathrm{f}} \circ F\left(S, S_{\mathrm{f}}\right)(x)\left|\overline{T}^t\right. = \overline{0}^t.$$

Thus, we know

$$\overline{y}^t = \left[S(x) + S \circ S_{\mathrm{f}}(y)\right]\left|\overline{T}^t\right. = S(x)\left|\overline{T}^t\right. + S \circ S_{\mathrm{f}}(y)\left|\overline{T}^t\right. = \overline{0}^t + \overline{0}^t = \overline{0}^t.$$

This end means that $F(S, S_{\mathrm{f}})$ is a causal system. QED

1.3.3 Time-Invariable Realization

Assume that $S : X \to Y$ is a linear functional time system. If the system S satisfies

$$\forall t \in T \forall x \in X \left(\lambda^t S\left(0^t \circ \sigma^t(x)\right) = S(x), \right. \tag{1.5}$$

then it is referred to as time invariably realizable, where the relation λ^t is defined by $\lambda^t(\bullet) = \sigma^t(\bullet|T_t)$.

Theorem 1.5 *Assume that $S : X \to Y$ is a linear functional time system. Then a sufficient and necessary condition for S to be time invariably realizable is that*

$$\forall t \in T \forall x \in X \left(S\left(0^t \circ \sigma^t(x)\right)|T_t = \sigma^t S(x)\right). \tag{1.6}$$

Proof ($\Rightarrow$) For this part of necessity argument, the assumption that S is time invariably realizable implies that the system S satisfies Eq. (1.5). Next, we show that Eq. (1.6) can be derivable from Eq. (1.5). To achieve this end, we have

$$\forall t \in T \forall x \in X, \quad \lambda^t S\left(0^t \circ \sigma^t(x)\right) = \lambda^t(y) = \sigma^{-t}\left(y\left|\overline{T}^t\right.\right) = S(x), \tag{1.7}$$

where $=S(0^t \circ \sigma^t(x)$. Therefore, we obtain

$$\sigma^t \circ \sigma^{-t}\left(y\left|\overline{T}^t\right.\right) = y\,|T_t = S\left(0^t \circ \sigma^t(x)\right)|T_t = \sigma^t S(x).$$

($\Leftarrow$) For this part of sufficiency argument, assume for any $t \in T$ and any $x \in X$ system S satisfies the condition in Eq. (1.6). Then, we have

$$\sigma^{-t} S\left(0^t \circ \sigma^t(x)\right)\left|\overline{T}_t\right. = \sigma^{-t}\sigma^t \mathrm{S}(x) = S(x). \tag{1.8}$$

Comparing Eqs. (1.5) and (1.8) leads to the conclusion that the given linear functional time system $S : X \to Y$ is time invariably realizable. QED

Theorem 1.6 Assume that *linear functional time systems $S : X \to Y$ and $S_f : Y \to X$ are time invariably realizable. Then a sufficient and necessity condition for the feedback system $F(S, S_f)$* to be *time invariably realizable is that*

$$\forall t \in T x \in D\left(F\left(S, S_f\right)\right) \ni S \circ S_f \circ F\left(S, S_f\right)\left(0^t \circ \sigma^t(x)\right)|T_t = \sigma^t S \circ S_f \circ F\left(S, S_f\right)(x), \tag{1.9}$$

Speaking differently, the fact that the feedback system $F(S, S_f)$ is time invariably realizable is equivalent to that $S \circ S_f \circ F(S, S_f)$ is time invariably realizable.

Proof (⇐) For the part of sufficiency argument, the assumption that $S \circ S_f \circ F(S, S_f)$ is time invariably realizable implies that for any $t \in T$ and any $x \in D(F(S, S_f))$, Eq. (1.9) is true. Hence, Theorem 1.5 implies that Eq. (1.6) is true. So, we have

$$S\left(0^t \circ \sigma^t(x)\right)|T_t + S \circ S_f\left(F(S, S_f)\left(0^t \circ \sigma^t(x)\right)|T_t\right)$$
$$= \sigma^t S(x) + \sigma^t S \circ S_f \circ F(S, S_f)(x),$$

which means

$$S\left[0^t \circ \sigma^t(x) + S_f\left(F(S, S_f)\left(0^t \circ \sigma^t(x)\right)\right)\right]\Big| T_t = \sigma^t S\left(x + S_f \circ F(S, S_f)(x)\right).$$

This end is equivalent to

$$F(S, S_f)\left(0^t \circ \sigma^t(x)\right)|T_t = \sigma^t F(S, S_f)(x).$$

Hence, Theorem 1.5 implies that $F(S, S_f)$ is time invariably realizable.

(⇒) For this part of the necessity argument, $F(S, S_f)$ is assumed to be time invariably realizable. So, Theorem 1.5 implies that

$$\forall t \in T \forall x \in D(F(S, S_f))\ \ F(S, S_f)\left(0^t \circ \sigma^t(x)\right)|T_t = \sigma^t F(S, S_f)(x). \tag{1.10}$$

By letting $F(S, S_f)(0^t \circ \sigma^t(x)) = y$ and $F(S, S_f)(x) = y'$, Eq. (1.10) implies

$$S\left(0^t \circ \sigma^t(x) + S_f(y)\right)|T_t = \sigma^t S\left(x + S_f\left(y'\right)\right),$$

and

$$S\left(0^t \circ \sigma^t(x)\right)|T_t + S \circ S_f(y)|T_t = \sigma^t S(x) + \sigma^t S \circ S_f\left(y'\right). \tag{1.11}$$

Both the time-invariable realizability of S and Theorem 1.5 indicate that Eq. (1.11) can be restated as follows:

$$S \circ S_f(y)|T_t = \sigma^t S\left(S_f\left(y'\right)\right). \tag{1.12}$$

Because Eq. (1.12) is the same as Eq. (1.9), we complete the proof. QED

1.3.4 Chaos and Attractor Under the Effect of Feedback Mechanism

Let S be an input-output system, satisfying $\varnothing \neq S \subset X \times Y$. Define $Z = X \cup Y$. Then S is a binary relation on Z and a subset $D \subset Z$, satisfying $D^2 \cap S = \varnothing$, is said to

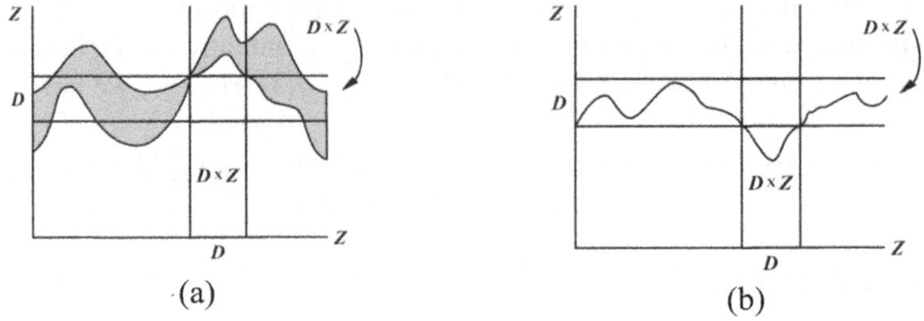

Fig. 1.4 The geometry of an attractor. (**a**) D is an attractor of S, and $S(x)$ contains at least one element for each x in $Z - D$. (**b**) D is an attractor of S, and $S : Z \to Z$ is a function

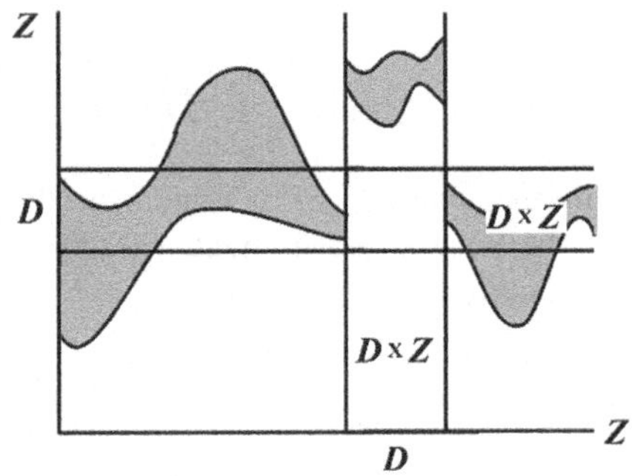

Fig. 1.5 D is a strange attractor of the system S.

be a chaos of S (Forrest, 2010; Zhu & Wu, 1987). Intuitively, D is seen as a chaos because system S has no control over the elements in D.

A subset $D \subset Z$ is known as an attractor of S (Forrest, 2010; Zhu & Wu, 1987) if for each $x \in Z - D$, $S(x) \cap D \neq \varnothing$. Figure 1.4a, b show the geometry of the concept of attractors. When S is not a function, Fig. 1.4a shows that the graph of S outside the vertical bar $D \times Z$ overlaps the horizontal bar $Z \times D$. When S is a function, Fig. 1.4b shows that the graph of S outside the vertical bar $D \times Z$ must be contained in the horizontal region $Z \times D$.

A subset $D \subset Z$ is said to be a strange attractor of system S (Forrest, 2010; Zhu & Wu, 1987), if D is both a chaos and an attractor of S. Figure 1.5 shows the case when a subset D of Z is a strange attractor of an input-output system S, where the square $D \times D$ is the only portion of the band $Z \times D$ over which the graph of S does not touch.

Theorem 1.7 *As long as $S \subset X \times X$ is a linear system, there must be a linear feedback component system $S_f : X \to X$ so that the following are equivalent:*

- *A subset $D \subset X$ is a chaos, or an attractor, or a strange attractor of system S; and.*
- *The subset D is a chaos, or an attractor, or a strange attractor of $F(S, S_f)$, correspondingly.*

Proof Evidently, no matter how a linear system $S \subset X \times X$ is defined, the desired learn feedback component system $S_f : X \to X$ can be defined as follows:

$$S_f(x) = 0_x, \quad \forall x \in X.$$

To finish the proof, we only need to check that $S = F(S, S_f)$. QED

Theorem 1.8 *Let $S \subset X \times X$ be a linear system, $S_f : X \to X$ a functional linear system and $D \subset X$, satisfying*

$$(X - D) \pm S_f(D) \subset X - D. \tag{1.13}$$

Then D is a chaos of S if and only if D is a chaos of the feedback system $F(S, S_f)$.

Proof ($\Rightarrow$) For this part of the necessity argument, suppose that $D \subset X$ is a chaos of S while not a chaos of the feedback system $F(S, S_f)$. Then for $(x, y) \in D^2 \cap F(S, S_f)$, the following holds true:

$$(x + S_f(y), y) \in S.$$

So, we have

$$x' = x + S_f(y) \notin D \quad \text{and} \quad x = x' - S_f(y).$$

This end contradicts Eq. (1.13), which implies that D must be a chaos of $F(S, S_f)$.

($\Leftarrow$) For this part of the sufficiency argument, the assumption that $D \subset X$ is a chaos of $F(S, S_f)$ but not a chaos of S implies that if $(x, y) \in D^2 \cap S$, $(x - S_f(y), y) \in F(S, S_f)$ holds true. That implies that $x' = x - S_f(y) \notin D$. This last equation is equivalent to $x = x' + S_f(y)$, which is contradictory to Eq. (1.13). Therefore, we can conclude that D must be a chaos of S. QED

Theorem 1.9 *Let $S \subset X \times X$ be a linear system and $S_f : X \to X$ a functional linear system, satisfying the following two conditions:*

$$X - D \subset \{x - S_f(y) : (x, y) \in S \cap ((X - D) \times D)\}, \tag{1.14}$$

and

$$X - D \subset \{x + S_f(y) : (x, y) \in F(S, S_f) \cap ((X - D) \times D)\}. \tag{1.15}$$

Then, a sufficient and necessary condition for $D \subset X$ to be an attractor of S is that D is an attractor of the feedback system $F(S, S_f)$.

Proof ($\Rightarrow$) For this part of the necessity argument, the assumption that $D \subset X$, satisfying Eq. (1.14), is an attractor of S but not of $F(S, S_f)$ implies that

$$\exists x \in X - D \ni F(S, S_f)(x) \cap D = \varnothing.$$

Equation (1.14) implies that

$$\exists (z, y) \in S \cap ((X - D) \times D) \ni z - S_f(y) = x.$$

Thus, we have $(x, y) \in F(S, S_f)$. That is, we have $y \in F(S, S_f)(x) \cap D$, a contradiction.

($\Leftarrow$) For this part of the sufficiency argument, assume that $D \subset X$, satisfying Eq. (1.15), is an attractor of $F(S, S_f)$, but not of S. Then, we have

$$\exists x \in X - D \ni S(x) \cap D = \varnothing.$$

At the same time, Eq. (1.15) implies that

$$\exists (z, y) \in F(S, S_f) \cap ((X - D) \times D) \ni z + S_f(y) = x.$$

Therefore, $(x, y) \in S$. That is, we have $y \in S(x) \cap D$, a contradiction. QED

1.4 When Government Economic Policies Become Effective: An Application

As an application of the abstract theory developed above, this section addresses the question of when government economic policies are expected to practically work in terms of economic development.

1.4.1 The Importance of the Government

First, let us look at how an economy, be it national, regional, or local, can be seen as a system. To do this, we first need to understand system as such a notion that models an organization or a structure (Klir, 2001) as a collection of some components or objects and a collection of relations that associate the objects in certain specified ways (Forrest, 2010). Hence, systems considered above represent special cases of this general understanding so that theoretically important and practically significant results can be derived. Symbolically, this general concept of systems can be written as follows: a system is an ordered pair $S = (M, R)$ of two sets, where the set M

contains all the isolated objects or elements of system S and the set R demonstrates all the particular relations that connect the elements in M into a whole (Lin, 1999).

With this understanding, it becomes evident that most business entities can be seen as systems. For example, each business organization consists of such components as employees, real-estate properties, equipment, and others, and a set of relations that associate the components into the recognizable organization. As for the economy of a nation, it can be theoretically modeled as an abstract system with all economic agents as components and associations that join the components in ways that are specific to the nation into a functional economy. The components of this systemic modeling include, but not limited to, consumers, families, business firms, levels of the government, various kinds of industries, economic sectors, and markets. Because each nation has its exclusively specific economy, it can be imagined that other than the fact that each nation has its specific set of components, these components are associated with each other in a nation-particular way. These associations describe how economic factors, such as money, information, knowledge, and talent, flow within the economy, and how consumer preferences and tastes evolve relentlessly. That is, the nation's economy can be modeled as a functional, hierarchical structure or a multi-leveled system. The operation of this system is guaranteed by the government through implementing various tools, such as policies, regulations, judicial procedures, law reinforcement, and the military might. The hierarchical structure of a nation's economy consists of leveled systems such as families, business firms, economic sectors, markets, and governments of various levels.

Speaking in terms of centralized systems (Forrest, 2010; Hall & Fagen, 1956; Lin, 1999), the uppermost level of the government functions as the center of the economy. Each change at this level, although only slightly, affects most areas of the national economy. The formal, systemic structure of the economy can be envisioned as a multi-dimensional hierarchy. The bottom of the hierarchy consists of individual consumers, whose consumptions, either for the purpose of basic living or that of enjoying a luxury lifestyle, mobilize all other layers of the economy. The middle of the hierarchy is occupied by business firms that respectively serve their particular segments of the totality of consumers. The layer second to the uppermost level of the hierarchy appears in the form of various markets, through which exchanges of information, knowledge, money, products, services, etc., take place.

This discussion indicates that the economy of a nation can be investigated as a multi-dimensional, leveled, centralized, large-scale complex system. Hence, the existence theorem of a centralized partial system (Forrest et al., 2020, p. 138) implies the following conclusion:

> Although there might be areas within a nation's economy over which the government has either no or little influence, the government can still exert its influence over at least one portion of the economy that is of roughly the same magnitude as the entire economy.

That is, the government not only provides an orientation for guiding principles on how the nation's economy should evolve, but also affects the majority of economic components, such as companies, industries, markets, etc. Recent examples of government-directed orientations of economic development include United States' Advanced Manufacturing Partnership (PCAST, 2012), Germany's Industry 4.0 (Industrie 4.0, 2011), India's "Make in India" (Business Standard, 2014), Japan's ultra-smart Society 5.0 (Government of Japan, 2015), China's "Made in China 2025" (State Council of the PRC, 2015), and United Kingdom's push for growing the Artificial Intelligence Industry (Hall & Pesenti, 2017). That is, a nation's government plays a non-negligible role in that nation's economic development. In other words, when a nation's government desires to promote the growth of the economy, it can readily employ policy tools to increase the degree of marketization, deepen political reform, and encourage dramatic rise of private enterprises or some other aspects of the economy. Such effort that is initiated by the government is most likely to achieve success due to the government's particular position as the center of the economy and the society. The reason why this approach of employing policy tools can indeed practically work is because the government can effectually levy its will on a large segment, although not the entire economy, of the economy, which in turn encourages economic agents to react accordingly—a feedback mechanism. Speaking differently, our discussion above leads to the following observation, for relevant studies, see Chemers (2001) and Kouzes and Posner (2007):

> Each commitment of the government represents a course of social influence and relevant sequence of supports. Within such a process, government officials are able to solicit a large number of economic agents to offer their assistance so that a predetermined national objective can be more or less accomplished.

If a system has a focus on reaching a determined collection of objectives and consequences, then the system is known as goal orientated (Klir, 2001). By combining the concept of goal-oriented systems, the systemic study of leadership (Lin & Forrest, 2011) and the discussions above, the conclusion below follows:

> The organizational structure of a nation can be modelled as a goal-oriented system, from which the importance of the government is shown in its capability to adjust the system's orientation. To align with such capability of the government, as a part of the economic feedback mechanism, individual economic agents alter and fine-tune their systemic orientations and operations.

1.4.2 When Government Economic Policies Actually Work

By continuing the discussion of the previous subsection and Sect. 1.2.2, this subsection studies the following question: under what conditions can the government of a nation apply policy tools to help with its nation's economic growth? In terms of the literature, some studies confirm that government policies are positively necessary for stimulating economic development (Andreoni & Chang, 2019; Howell et al., 2019), while others maintain that such policy tools do not really work (Hemphill, 2014; Jomo, 2019). This section fills this gap in the literature by resolving this dichotomy through showing the fact that the usefulness and effectiveness of government policies is conditional.

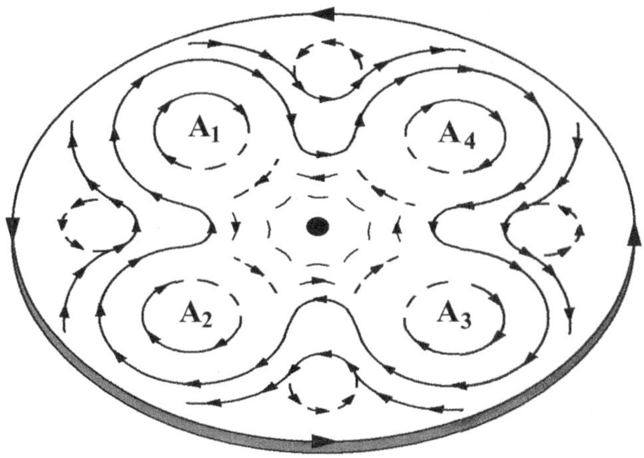

Fig. 1.6 The appearance of a goal-oriented system

Assume that through using policy tools, the nation of concern can accomplish such an objective that potentially benefit a large number of economic agents. Then the concepts of centralized and feedback systems (Forrest, 2010; Forrest et al., 2020, p. 138; Hall & Fagen, 1956; Lin, 1999) and discussions in the previous subsection collectively imply the following fact:

> The government of the nation can use economic policies to obtain public support by exerting its organizational influence and formulate such a national ambition that will involve a major portion of the economy.

For a relevant but different conclusion, see Chemers (2001).

To assist with our discussion, let us depict the economy of a nation as a spinning dish (Fig. 1.6), as studied in the well-known dishpan experiment (Fultz et al., 1959). In particular, the experiment is composed of a spinning dishpan that is filled with fluid. Corresponding to our discussion here, this dish models the nation's economy, while the moving "fluid" represents the movement of such economic factors as money, information, knowledge, goods, services, and others within the economy. As the dish spins at a sufficiently fast speed, the fluid demonstrates the flow pattern given in Fig. 1.6 due to uneven distributions of natural resources and human talents (Forrest et al., 2020). In other words, the chaotic flow pattern reflects the fact that imbalances in economic development of geographical regions represent a commonly existing phenomena, while the black dot in the center models the social and political position of the government.

One noteworthy characteristic of Fig. 1.6 is that the regional fields located in between A_1, A_2, A_3, and A_4, swirl in opposite directions when compared to the spin directions of A_1, A_2, A_3, and A_4. Such differences in spin directions intuitively demonstrate the possibility that business firms that aim at achieving different or even conflicting goals can peacefully coexist and can help strengthen each other economically. That is, the following conclusion holds true:

> If a nation's government is able to execute its adopted policies effectively, then it is capable of forming a united front of effort for a large magnitude of individual economic agents.

Speaking differently, this conclusion indicates that although individual enterprises may very well aim at achieving their respectively different and often conflicting business goals, the government's ability to implement its policies effectively will help line up the objectives of a large segment of the economy, even if these objectives might be inconsistent or conflicting with each other.

Collectively, the previous conclusions, established on the basis of systemic thinking, naturally lead to the result below:

> If a nation's government aims at advancing the national economy, then it can accomplish this initial ambition by adopting and implementing appropriate policies that potentially benefit a majority segment of the entire economy.

To summarize what are obtained in this section, for an adopted and implemented policy to produce a desired outcome, the essence is that the policy has to aim at potentially benefiting a large segment of the national economy. With such an unwavering aim, although conceptual and practical specifications may be different from the implementation of one policy to another, each implemented policy will provide a common goal for a large percentage of economic agents to target at. Over time, the unwavering aim of the government will become a bright highlight for the nation's citizens to feel enthusiastic about and for various economic agents to orient their efforts. For related studies, see Forrest and Liu (to appear).

1.5 A Few Final Words

As suggested by the title, this chapter studies how properties of MT time systems are affected by the feedback mechanism. To this end, nine theorems are developed on the basis of set theory, as strongly advocated by Professors Mihajlo Mesarovic and Yasuhiko Takahara since the 1960s (Mesarovic, 1964). These theorems establish conditions under which characteristics of MT time systems, such as linearity, strong stationality, (pre-)causality, time-invariable realizability, chaos, attractors, and strange attractors, are maintained under the effect of the feedback mechanism. Many of them provide necessary and sufficient conditions for each of these properties to be preserved.

Other than the aforementioned theoretical results that help enrich what has been advanced by Professor Yasuhiko Takahara and his school of thoughts, this work also contributes to the literature of economics in two different ways. First, by demonstrating how the feedback mechanism organically associates the quality of manufacturing and the depth/width of the market exchanges within a nation's economy, we are able to show the importance of manufacturing in the economic development of a nation. This end helps settle an inconsistent literature on whether or not manufacturing had historically played an important role in a nation's industrialization. For example, many scholars (e.g., Merati et al., 2019) believe so through their case-specific studies, while many others (e.g., Szirmai & Verspagen, 2015) maintain that the opposite is true.

Secondly, this work fills a gap in the literature on economic (industrial) policies in terms of whether or not such policies actually work in real life as imagined and desired. Instead of participating in the endless rounds of relevant debates (Andreoni & Chang, 2019), this chapter develops conditions under which an economic policy will play its expected roles.

References

Andreoni, A., & Chang, H. J. (2019). The political economy of industrial policy. *Structural Change and Economic Dynamics, 48*, 136–150.

Bayliss, L. E. (1966). Living control systems. San Francisco.

Business Standard. (2014). *Focus on 'Make in India—Ministry of Heavy Industries & Public Enterprises*. Retrieved April 28, 2021, from https://www.business-standard.com/article/government-press-release/focus-on-make-in-india-114092501206_1.html

Chemers, M. M. (2001). Cognitive, social, and emotional intelligence of transformational leadership: Efficacy and effectiveness. In R. E. Riggio, S. E. Murphy, & F. J. Pirozzolo (Eds.), *Multiple intelligences and leadership* (pp. 139–160). Lawrence Erlbaum.

Deng, J.-L. (1985). *Grey control systems*. Hua-zhong Institute of Technology Press.

Forrest, J. Y. L. (2010). *General systems theory: Foundation, intuition and applications in business decision making*. Springer.

Forrest, J. Y. L., Buttermore, J., & Wajda, T. A. (2017). At Nash equilibrium when new market competitions appear? *Kybernetes, The International of Cybernetics, Systems and Management Science, 46*(2), 256–271.

Forrest, J. Y. L., & Liu, Y. (to appear). *Value in business—A holistic, systems-based approach to creating and capturing value*. Springer.

Forrest, J. Y. L., Nicholls, J., Schimmel, K., & Liu, S. F. (2020). *Managerial decision making: A holistic approach*. Springer.

Forrest, J., Ying, Y. R., & Gong, Z. W. (2018a). *Currency wars: Offense and defense through systemic thinkings*. CRC Press.

Forrest, J. Y. L., Zhao, H. C., & Shao, L. (2018b). Engineering rapid industrial revolutions for impoverished agrarian nations. *Theoretical Economics Letters, 8*, 2594–2640. https://doi.org/10.4236/tel.2018.811166

Fultz, D., Long, R. R., Owens, G. V., Bohan, W., Kaylor, R., & Weil, J. (1959). *Studies of thermal convection in a rotating cylinder with some implications for large-scale atmospheric motion* (Meteorological monographs (American Meteorological Society) (Vol. 21, no. 4)). Springer.

Government of Japan. (2015). Retrieved April 16, 2019, from https://www8.cao.go.jp/cstp/kihonkeikaku/5basicplan_en.pdf

Hall, A. D., & Fagen, R. E. (1956). Definitions of systems. *General Systems, 1*, 18–28.

Hall, W., & Pesenti, J. (2017). *Growing the artificial intelligence industry in the UK*. Retrieved October 15, 2017, from https://assets.publishing.service.gov.uk/government/uploads/system/uploads/attachment_data/file/652097/Growing_the_artificial_intelligence_industry_in_the_UK.pdf

Heaton, H. (2017). Industrial revolution. In: *Encyclopedia of the social sciences* (Vol. VIII (1033)). In R. M. Hartwell (Ed.) (2017), *The causes of the industrial revolution in England* (pp. 30–52). Routledge.

Hemphill, T. A. (2014). The US advanced manufacturing initiative: Will it be implemented as an innovation—Or industrial—Policy? *Innovation: Organization & Management, 16*(1), 67–70.
Henig, M. I. (1983). Vector-valued dynamic programming. *SIAM Journal of Control and Optimization, 3*, 237–245.
Howell, T. R., Noellert, W. A., MacLaughlin, J. H., & Wolff, A. W. (2019). *The microelectronics race: The impact of government policy on international competition*. Routledge.
Industrie 4.0. (2011). *BMBF-Internetredaktion*. Retrieved April 28, 2021, from https://www.bmbf.de/de/zukunftsprojekt-industrie-4-0-848.html
Jomo, K. S. (2019). *Southeast Asia's misunderstood miracle: Industrial policy and economic development in Thailand, Malaysia and Indonesia*. Routledge.
Klir, G. (2001). *Facets of systems science*. Springer.
Kouzes, J., & Posner, B. (2007). *The leadership challenge*. Jossey Bass.
Lin, Y. (1989). Some properties of linear systems. *International Journal of Systems Sciences, 20*(6), 927–937.
Lin, Y. (1999). *General systems theory: A mathematical approach*. Plenum and Kluwer Academic Publishers.
Lin, Y., & Forrest, B. (2011). *Systemic structure behind human organizations: From civilizations to individuals*. Springer.
Lin, Y., & Ma, Y. (1987). Some properties of linked time systems. *International Journal of General Systems, 13*(2), 125–134.
Lin, Y., & Ma, Y. H. (1990). General feedback systems. *International Journal of General Systems, 18*(2), 143–154.
Ma, Y., & Lin, Y. (1992). A representation theorem of linear I/O systems. *Systems Research, 19*(2), 42–48.
Merati, M., Salehi, V., & Rafiei, A. (2019). A unique mathematical programming algorithm for performance optimization of organizational indicators in manufacturing sector. *Journal of Industrial and Production Engineering, 36*(7), 461–474.
Mesarovic, M. D. (1964). Views on general systems theory. In M. D. Mesarovic (Ed.), *Proceedings of the 2nd Systems Symposium at Case Institute of Technology*. Wiley.
Mesarovic, M. D., & Takahara, Y. (1975). *General systems theory: Mathematical foundations*. Academic Press.
Mesarovic, M. D., & Takahara, Y. (1989). *Abstract systems theory*. Springer-Verlag.
Milsum, J. H. (1966). *Biological control systems analysis*. McGraw-Hill.
Negoita, C. V. (1992). Pullback versus feedback. In C. V. Negoita (Ed.), *Cybernetics and applied systems* (pp. 239–247). Marcel Dekker.
PCAST. (2012). *Report to the President on capturing domestic competitive advantage in advanced manufacturing*. Retrieved July, 2012, from https://permanent.access.gpo.gov/gpo34116/pcast_amp_steering_committee_report_final_july_17_2012.pdf
Porter, M. E. (1985). *Competitive advantage: Creating and sustaining superior performance*. Free Press.
Rostow, W. W. (1960). *The stages of economic growth: A non-communist manifesto*. Cambridge University Press.
Saito, T. (1986). Feedback transformation of linear systems. *International Journal of Systems Science, 17*(9), 1305–1315.
Saito, T. (1987). Structure of the class of linear functional time systems under output feedback transformation. *International Journal of Systems Science, 18*(3), 1017–1027.
Saito, T., & Mesarovic, M. D. (1985). A meaning of the decoupling by feedback of linear functional time systems. *International Journal of General Systems, 11*(1), 47–61.
State Council of the PRC. (2015). *Notice of State Council of the PRC on printing and distributing "Made in China 2025"*. Retrieved May 8, 2015, from http://www.gov.cn/zhengce/content/2015-05/19/content_9784.htm
Szirmai, A., & Verspagen, B. (2015). Manufacturing and economic growth in developing countries, 1950-2005. *Structural Change and Economic Dynamics, 34*, 46–59.

Takahara, Y., & Asahi, T. (1985). Causality invariance of general systems under output feedback transformations. *International Journal of Systems Science, 16*(10), 1185–1206.

Wen, Y. (2016). *The making of an economic superpower: Unlocking China's secret of rapid industrialization*. World Scientific.

Wonham, W. M. (1979). *Linear multi-variable control: A geometric approach*. Springer.

Wu, X. M. (1981). Pansystems analysis—A new exploration of interdisciplinary investigation (with appendices on discussions on medicine, chemistry, logic and economics). *Scientific Exploration, 1*(1), 125–164.

Zadeh, L. (1965). Fuzzy sets. *Information and Control, 8*(3), 338–353.

Zhu, X. D., & Wu, X. M. (1987). An approach on fixed pansystems theorems: Panchaos and strange panattractor. *Applied Mathematical Mechanics, 8*(4), 339–344.

Chapter 2
Generalization of Law of Requisite Variety

Kyoichi Kijima

Abstract The purpose of the present research is to generalize and formulate Ashby's Law of Requisite Variety in the context of a goal seeking system and its relevant environment in the framework of logico-mathematical general systems theory.

The law, one of the most well-known laws in the general systems theory, intuitively claims that "only the variety can destroy the variety." The claim is so simple and has attracted more and more attention from a wide range of disciplines including organization theory, organizational cybernetics, and control theory. In this article, in order to obtain deeper and rigid implications and insights of the law, by introducing "subjective external variety" of the "dynamic environment," we will argue the balancing law concerning the relationship between a goal seeking system and its relevant environment. In this research we will also clarify the meaning of the internal model principle, which says that in order to control the environment the system should incorporate an internal model of it.

Keywords Mathematical general systems theory · Goal seeking system · Environment · Variety · Ashby's law of requisite variety · Decision making · Internal model principle

2.1 Introduction

General Systems Theory has developed at least two streams of research: General Theory of Systems and Theory of General Systems. General Theory of Systems tries to develop general theory by investigating some common or universal properties

---

An earlier version of this article was published as Kijima et al. (1986).

K. Kijima (✉)
School of Business Management, Bandung Institute of Technology, Bandung, Indonesia
e-mail: kijima@kijima-lab.com

K. Kijima et al. (eds.), *Systems Research I*, Translational Systems Sciences 26,
https://doi.org/10.1007/978-981-16-9937-5_2

of all systems. This direction of the system research was clearly shown when the Society for General Systems Research (SGSR) was founded in 1954 by Ludwig von Bertalanffy, Kenneth Boulding, Ralph Gerard, and Anatol Rapoport. (The society adopted its current name, the Society for Systems Sciences (ISSS), in 1988 to reflect its broadening scope) (ISSS 2021).

The initial purpose of the society was "to encourage the development of theoretical systems that are applicable to more than one of the traditional departments of knowledge," with the following principal aims:

- To investigate the isomorphy of concepts, laws, and models in various fields, and to help in useful transfers from one field to another.
- To encourage the development of adequate theoretical models in areas that lack them.
- To eliminate the duplication of theoretical efforts in different fields.
- To promote the unity of science through improving the communication among specialists.

Indeed, toward this direction not only the founders but also other great researchers including James Miller developed tremendous contributions. Among others Ross Ashby's Law of Requisite Variety is certainly one of the most famous principles advocated in the field (Ashby, 1963).

The other stream of General Systems Theory is Theory of General Systems, which tries to focus on "specific systems called general systems" and analyze their characteristics in a rigid mathematical way. This direction, often called logico-mathematical systems theory, was pioneered by Mihajlo Mesarovic, Yasuhiko Takahara, A. Wayne Wymore, George Klir, and so on (Klir, 2001).

This article aims to generalize and formulate the Law of Requisite Variety in the framework of logico-mathematical general systems theory, so that it is positioned at the intersection of the both in the sense that it investigates the topic from general theory of systems by approach of theory of general systems (Kijima, 1983, 1984; Kijima et al., 1985; Nakano, 1983).

Though the Law of Requisite Variety has many forms, its essence is very simple: "only variety can destroy variety." The basic logic of the law is that control can be obtained only if the variety of the controller is at least as great as the variety of the environment to be controlled.

The simple claim has given impacts on various disciplines including organization theory, social systems theory, and organizational cybernetics (Kickert, 1980). For instance, Stafford Beer adopts it as the basis for his concept of viable system model (VSM) in organizational management (Beer, 1981). The law has come to be understood as a simple proposition: If a system is to be able to deal successfully with the diversity of challenges that its environment produces, then it needs to have a repertoire of responses, which is (at least) as nuanced as the problems thrown up by the environment. Beer proposed the concept of variety engineering to argue two alternatives for a controller to attain balance between its internal variety and environment's external variety, i.e., variety reduction and variety amplification. The former means the controller copes with environmental challenges by reducing the variety of the environment. However, if variety reduction seems unfeasible, the controller is no longer viable so that they have to find other ways by increasing or amplifying their variety.

In the balancing law of internal and external variety, the key word "variety" is defined as the number of the distinguishable items (or distinguishable states of some item). However, although this kind of verbal definitions seems easily understandable at a first glance, some ambiguity about the basic definitions seems to remain.

Mesarovic claims two distinct approaches to the representation of a system: The "terminal approach" and the "goal seeking approach" (Mesarovic, 1967). The terminal approach is the conventional representation of a system as an entity that looks at a system from the outside and defines it in terms of causality (cause–effect relationship), as is done in physics, chemistry, engineering, etc. On the other hand, the goal seeking approach looks at a system as a white box focusing on the system's goals and goal seeking behavior in terms of means–ends relationship.

The present research generalizes and formulates the key concepts of the law, namely, external and internal variety, in the context of relationship between a goal seeking system and its environment in the framework of logico-mathematical general systems theory (Kijima et al., 1986).

To achieve the purpose we, taking the goal seeking approach, formulate a general balancing principle between the goal seeking system and its environment, at least two new attributes: First, we take into account dynamic behavior of environment. As pointed out above, while "external variety" of the environment is usually understood as the number of the distinguishable possible states of it, there have been no explicit discussions on dynamic state transition of the environment. Though, it is much more natural to assume that states of the environment take dynamic transition time to time. By arguing such a dynamic transition process of states of the environment we will also reveal the meaning of the internal model principle (Wonham, 1976), which says that in order to control the environment the system should incorporate an internal model of it.

Second, we introduce subjective variety. In the research so far variety is basically "objectively" defined as the number of possible states. For the case of environment's external variety, it means the number of possible states that the environment can take. However, this seems too restrictive. Let us suppose two companies or goal seeking systems belong to the same industry and face the same environment. It is quite natural that one company identifies the environment simple and easy to treat, while the other finds it complex and troublesome. It implies that their definition of variety should be different from each other. In order to argue this point explicitly, in this study, we extend the definition of variety to a subjective concept based on the relationship between the system and environment.

In this research we will make these two important extensions and develop rigorous and transparent discussion in the framework of logico-mathematical general systems theory.

2.2 Goal Seeking System and Environment

As mentioned in the previous section, in this research we generalize and formulate the law in terms of relationship between a goal seeking system (a controller) and its relevant environment (a controlled system) in the framework of logico-mathematical general systems theory (Fig. 2.1).

We represent the environment a goal seeking system faces by a pair $\mathcal{U} = (U, \alpha)$, where $U \subset \hat{U}$ is a nonempty set called an underlying set of the environment ($\hat{U}$ is a universe set of U), while α is a one-stage transition map of the environment, i.e., a generator of the environment. If an initial state of the environment is $u_0 \in U$, then α generates a sequence of $u_0, \alpha(u_0), \alpha^2(u_0)$, and so on.

Next, we formulate a goal seeking system in the following way. We assume that the goal seeking system knows the underlying set U of its relevant environment but has no knowledge about the dynamics of the environment (Miyazawa, 1983).

Hence, we represent a goal seeking system $\mathcal{S}$ by

$$\mathcal{S} = (\Phi, P, G, M, Y, U, \eta, d).$$

$M \subset \hat{M}$ and $Y \subset \hat{Y}$ are the sets of alternatives (or decision variables) and outcomes, respectively, where $\hat{M}$ and $\hat{Y}$ are universe sets of alternatives and outcomes, respectively. $P : M \times U \rightarrow Y$ and $G : M \times U \times Y \rightarrow R$ are called a process function and an objective function, respectively, where R denotes the set of all reals (Fig. 2.2). It is often convenient if we define a performance function $g : M \times U \rightarrow R$ by $g(m, u) = G(m, u, P(m, u))$.

Φ is called a "value system," which determines decision behaviors of the goal seeking system and usually a predicate containing $m \in M, u \in U, P$ and G. Under each $u \in U$ the value system Φ determines the set of solutions $\phi_{\mathcal{S}}(u)$ by

$$\phi_{\mathcal{S}}(u) = \{m \in M | m \text{ satisfies } \Phi(u)\}.$$

If no confusion arises, we will simply write $\phi(u)$ for $\phi_{\mathcal{S}}(u)$. ϕ is called a decision principle.

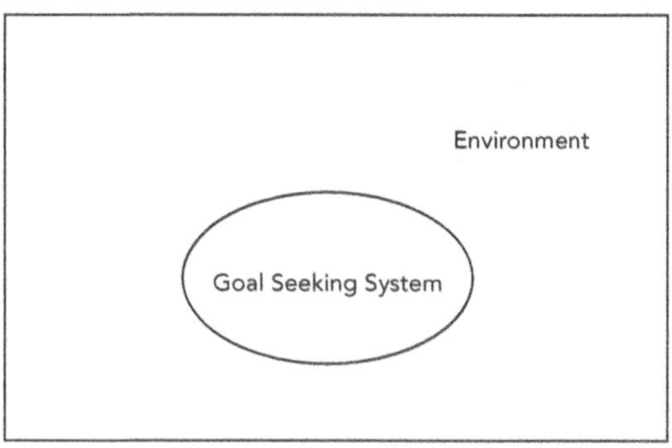

Fig. 2.1 Goal seeking system and its environment

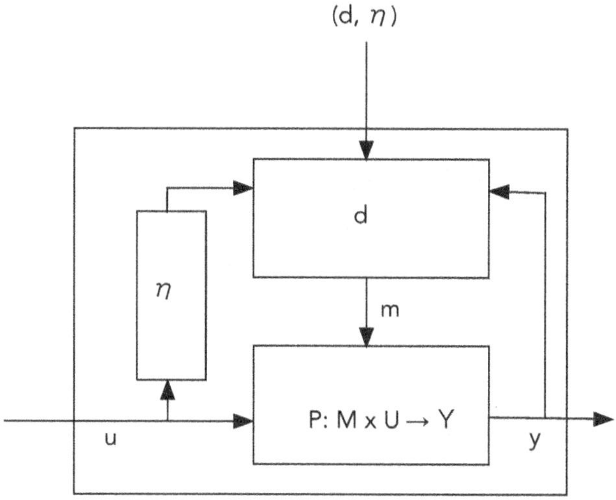

Fig. 2.2 Goal seeking system

A simple satisfactory decision principle,

$$\phi_1(u) = \{m \in M | g(m, u) \geq T(u)\},$$

where $T : U \rightarrow R$ is called an aspiration level, and an optimization principle

$$\phi_2(u) = \{m \in M | (\forall m' \in M)(g(m, u) \geq g(m', u))\}$$

are typical examples of the decision principle.

In this article the goal seeking system S is assumed to specify one and only one solution in M for every $u \in U$, that is, ϕ is a function such that $\phi : U \rightarrow M$ rather than a set function $\phi : U \rightarrow \wp(M)$, where $\wp(M)$ denotes the power set of M. It is because though there may be several alternatives in $\phi(u)$ for $u \in U$, often they can be identified with the same. This assumption looks quite sound.

$\eta : U \rightarrow V$ is an observation function that assigns an observation data $\eta(u)$ for every $u \in U$, where V be a set of observation data. For simplicity we assume η is surjective, i.e., $\eta(U) = V$. Then we can factorize η through $U/ker\ \eta$, where $U/ker\ \eta$ is the quotient set induced by $ker\ \eta \subset U \times U$ (Fig. 2.3). $ker\ \eta$ is an equivalence relation on U defined by

$$(\forall u, u' \in U)((u, u') \in ker\ \eta \Longleftrightarrow \eta(u) = \eta(u')).$$

For notational convenience we will simply write η for $ker\ \eta$. Since we have a bijection between V and U/η, we may identify the natural map $U \rightarrow U/\eta$ with the observation function $\eta : U \rightarrow V$ and can write $\eta : U \rightarrow U/\eta$.

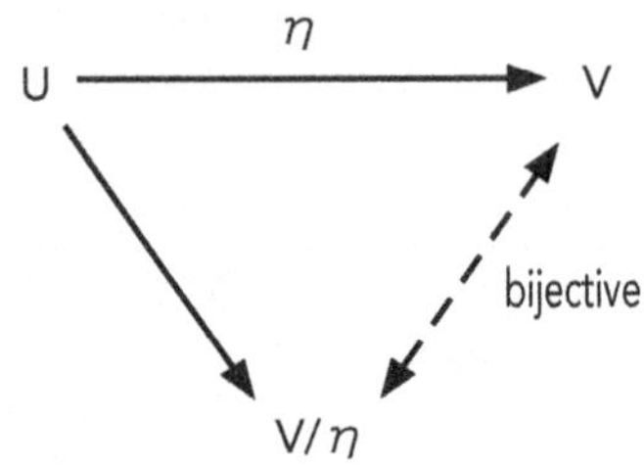

Fig. 2.3 Observation function

The equivalence relation η represents "accuracy" or "identification ability" of the observation η. That is, since for any u and u' in U $(u, u') \in \eta$ implies $\eta(u) = \eta(u')$, the system cannot distinguish u from u' by using the observation function η. As still be mentioned later, how to determine the accuracy of the observation function is deeply related with the variety the system possesses.

Finally, d is a decision function from U/η to M. d assigns an alternative $m \in M$ to each observed data $[u] \in U/\eta$ by

$$m = d([u]),$$

where $[u]$ denotes the equivalence class of u with respect to η. When U/η and M are given, we will represent the set of all decision functions by $hom(U/\eta, M)$, i.e.,

$$hom(U/\eta, M) = \{d | d : U/\eta \to M\}.$$

Based on consideration we will represent the class of goal seeking systems considered in the present article by

$$\tilde{\mathcal{S}} = \{\mathcal{S} = (\Phi, P, G, M, Y, U, \eta, d) | U \subset \hat{U}, M \subset \hat{M}, Y \subset \hat{Y},$$
$$P : M \times U \to Y, G : M \times U \times Y \to R \text{ and } \phi \text{ is a function of } U \text{ to } M\}.$$

2.3 Internal Model Principal and Anticipative Decision Making

Let $\mathcal{S} = (\Phi, P, G, M, Y, U, \eta, d)$ and $\mathcal{U} = (U, \alpha)$ be a goal seeking system and its relevant environment, respectively.

We begin with the following definition of system–environment consistency.

Definition 1 (System–Environment Consistency) We call a goal seeking system $\mathcal{S}$ and its relevant environment $\mathcal{U} = (U, \alpha)$ are consistent, if

$$\phi \subset \phi\alpha$$

holds, where for notational convenience, we write ϕ and $\phi\alpha$ for the equivalence relations $ker\ \phi$ and $ker\ \phi\alpha$, respectively.

Since the system–environment consistency implies that

$$(\forall u \in U)(\forall u' \in U)(\phi(u) = \phi(u') \to \phi\alpha(u) = \phi\alpha(u')),$$

ϕ is a congruence relation with respect to α. It implies that the environment does not change so "rapidly" with respect to the decision principle.

There are at least two ways for the system to make the assumption hold: The first is to proactively take actions toward the environment $\mathcal{U} = (U, \alpha)$ to change its dynamics α. The second is to modify the value system Φ, the objective function G, and/or the process function P.

Definition 2 (Initial Solvability) Let $\mathcal{S} = (\Phi, P, G, M, Y, U, \eta, d)$ and $\mathcal{U} = (U, \alpha)$ be a goal seeking system and its relevant environment, respectively. (η, d) is called to have initial solvability if

$$(\forall u \in U)(d\eta(u) = \phi(u))$$

holds.

This definition implies that if the system has initial solvability, then $d\eta$ can always "solve the problem" by finding a solution $\phi(u)$ whichever $u \in U$ occurs.

Now we have,

Proposition 1 *Let $\mathcal{S} = (\Phi, P, G, M, Y, U, \eta, d)$ and $\mathcal{U} = (U, \alpha)$ be a goal seeking system and its relevant environment, respectively. Suppose $\mathcal{S}$ and $\mathcal{U}$ are consistent. If (η, d) has initial solvability, then there is $\beta : M \to M$ such that*

$$(\forall u \in U)(\forall t \in \{0, 1, 2, \ldots\})(\phi\alpha^t(u) = \beta^t d\eta(u)).$$

In this case the system is called anticipative while $\beta : M \to M$ is called an internal model of the environment $\mathcal{U} = (U, \alpha)$ with respect to Φ (Refer to Fig. 2.4).

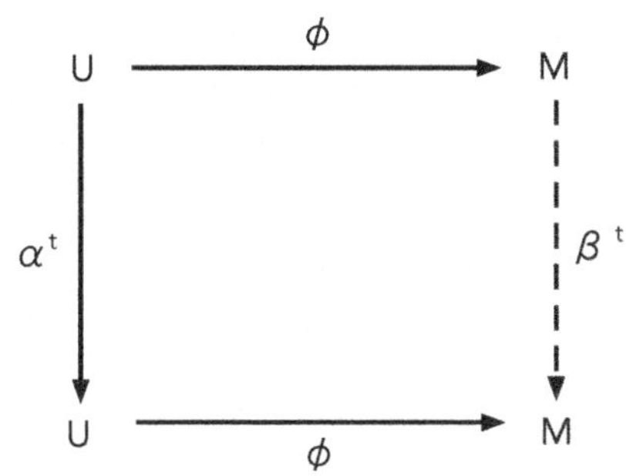

Fig. 2.4 Internal model

Proof Refer to Appendix 1.

The proposition says that under the conditions the system can "successfully manage" the environment at any time by incorporating an internal model β of α, if the system can manage the initial environment. If it is the case, the system is called anticipative because it can generate solutions anticipatively by using β as $\beta^t d\eta(u)$ for the future without actual observation $\alpha^t(u)$ of the environment.

In particular, if ϕ is surjective, then we will say the system is irredundant with respect to $(\Phi, \mathcal{U})$.

Corollary 2 *Let* $\mathcal{S} = (\Phi, P, G, M, Y, U, \eta, d)$ *and* $\mathcal{U} = (U, \alpha)$ *be a goal seeking system and its relevant environment, respectively. Suppose the same conditions as in Proposition 1 hold. Furthermore, assume the system is irredundant with respect to* $(\Phi, \mathcal{U})$. *Then the internal model is unique.*

Proof Refer to Appendix 2.

Conversely, if the system incorporates an internal model of the environment, then the system necessarily has initial solvability and it is consistent with the environment. That is,

Proposition 3 *Suppose the goal seeking system* $\mathcal{S} = (\Phi, P, G, M, Y, U, \eta, d)$ *is anticipative, i.e., it incorporates an internal model* β *of the environment* $\mathcal{U} = (U, \alpha)$ *with respect* Φ. *Then, the system has initial solvability and the system–environment consistency holds.*

Proof Refer to Appendix 3.

By combining Propositions 1 and 3 we have

Theorem 4 *Let* $\mathcal{S} = (\Phi, P, G, M, Y, U, \eta, d)$ *and* $\mathcal{U} = (U, \alpha)$ *be a goal seeking system and its relevant environment, respectively.* $\mathcal{S}$ *and* $\mathcal{U}$ *are consistent and* (η, d) *has initial solvability if and only if the system is anticipative.*

2.4 Subjective Varieties of Environment and System

Now we introduce the concept of "subjective" external and internal variety.

Definition 3 (External Variety) Let $\mathcal{S} = (\Phi, P, G, M, Y, U, \eta, d)$ and $\mathcal{U} = (U, \alpha)$ be a goal seeking system and its relevant environment, respectively. A binary relation $E(\mathcal{S}, \mathcal{U}) \subset U \times U$ is called the external variety of the environment $\mathcal{U}$ with respect to $\mathcal{S}$ if it is defined by

$$(\forall u, u' \in U)((u, u') \in E(\mathcal{S}, \mathcal{U}) \leftrightarrow (\forall t \in \{0, 1, 2, \ldots\})(\phi\alpha^t(u) = \phi\alpha^t(u')).$$

Lemma 5 *$E(\mathcal{S}, \mathcal{U})$ is an equivalence relation.*

Proof Clear.

Definition 3 claims that if u and $u' \in U$ are in $E(\mathcal{S}, \mathcal{U})$, then the solutions under $\alpha^t(u)$ and $\alpha^t(u')$ are the same for any t. We should notice that the equivalence relation depends not only on $\mathcal{U}$ but also on $\mathcal{S}$. Indeed, the external variety of the environment is defined as subjective to the system in the sense that it is affected by the "value system" Φ of $\mathcal{S}$ as well.

Example 1 If $E(\mathcal{S}, \mathcal{U}) = \Delta$ holds, where Δ denotes the diagonal, then we have $(u, u') \notin E(\mathcal{S}, \mathcal{U})$ for every u and u' in U whenever $u \neq u'$. Hence we have

$$(\exists t)(\phi\alpha^t(u) \neq \phi\alpha^t(u')).$$

It means that in the case of $E(\mathcal{S}, \mathcal{U}) = \Delta$ every state of U is essentially different with respect to $\mathcal{S}$.

Example 2 If $E(\mathcal{S}, \mathcal{U}) = U \times U$ holds, then we have $(u, u') \in E(\mathcal{S}, \mathcal{U})$ for every u and u' in U. Hence we have

$$(\forall t)(\phi\alpha^t(u) = \phi\alpha^t(u')).$$

It means that in the case of $E(\mathcal{S}, \mathcal{U}) = U \times U$ every state of U is essentially the same with respect to $\mathcal{S}$.

Now we turn to defining "internal" variety of the goal seeking system. It depends on the accuracy of the observation and the set of alternatives. That is,

Definition 4 (Internal Variety) Let $\mathcal{S} = (\Phi, P, G, M, Y, U, \eta, d)$ and $\mathcal{U} = (U, \alpha)$ be a goal seeking system and its relevant environment, respectively. Then $hom(U/\eta, M)$ is called the internal variety of $\mathcal{S}$ with respect to $(\Phi, \mathcal{U})$.

$hom(U/\eta, M)$ represents the set of all possible decision functions available for the goal seeking system. Hence we can see that when the size of the underlying set U of the environment is too large for the system to control, it has at least two ways to amplify the internal variety by increasing the cardinality of $hom(U/\eta, M)$; the first is to refine the equivalence relation η induced by the observation function, while the second is to extend the size of M. The former implies increase of observation accuracy, while the latter suggests generation of new alternatives. The definition of the internal variety of the system formulates our intuition explicitly.

Since the internal variety of the system is essentially related to the decision making behavior generated by Φ, we introduce optimality criterion for the decision functions.

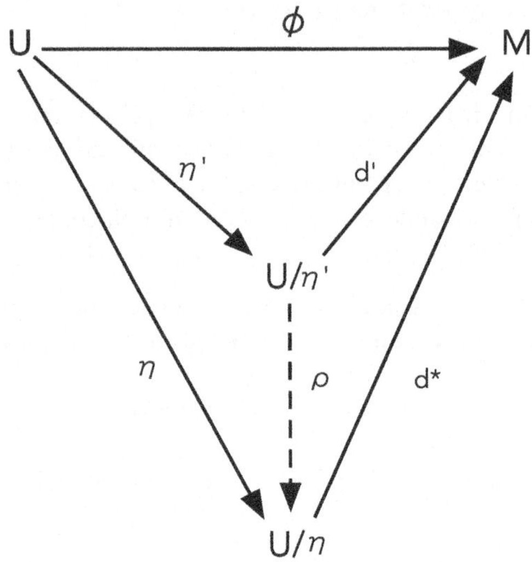

Fig. 2.5 Optimal decision function

Definition 5 (Optimal Decision Function) Let $\mathcal{S} = (\Phi, P, G, M, Y, U, \eta, d)$ and $\mathcal{U} = (U, \alpha)$ be a goal seeking system and its relevant environment, respectively. d^* is an optimal decision function with respective to $(\Phi, \mathcal{U})$ if and only if

1. (η, d^*) has initial solvability.
2. For every pair (η', d') which has initial solvability, there is a natural map $\rho : U/\eta' \to U/\eta$ such that Fig. 2.5 is commutative.

An optimal decision function "costs" the least among the decision functions that have initial solvability.

Whether or not the system can really choose an optimal decision function entirely depends on the specification of $hom(U/\eta, M)$. Indeed, if $hom(U/\eta, M)$ is not appropriate, all the decision functions in it may fail to be optimal. We, hence, make the following definition of optimality of $hom(U/\eta, M)$.

Definition 6 (Optimal Internal Variety) Let $\mathcal{S} = (\Phi, P, G, M, Y, U, \eta, d)$ and $\mathcal{U} = (U, \alpha)$ be a goal seeking system and its relevant environment, respectively. $hom(U/\eta, M)$ is said to be an optimal internal variety of the system $\mathcal{S}$ with respect to $(\Phi, \mathcal{U})$ if and only if $hom(U/\eta, M)$ contains an optimal decision function with respect to $(\Phi, \mathcal{U})$.

The following proposition is one of the main claims of this research.

Proposition 6 *Let $\mathcal{S} = (\Phi, P, G, M, Y, U, \eta, d)$ and $\mathcal{U} = (U, \alpha)$ be a goal seeking system and its relevant environment, respectively. $hom(U/\eta, M)$ is an optimal internal variety of the system $\mathcal{S}$ with respect to $(\Phi, \mathcal{U})$ if and only if $\eta = E(\mathcal{S}, \mathcal{U})$.*

Proof Refer to Appendix 4.

We can see that the condition $\eta = E(\mathcal{S}, \mathcal{U})$ claims η can observe the external variety in a mutually exclusive and collectively exhaustive (MECE) way.

An optimal internal variety has the following uniqueness property.

Lemma 7 *Let $\mathcal{S} = (\Phi, P, G, M, Y, U, \eta, d)$ and $\mathcal{U} = (U, \alpha)$ be a goal seeking system and its relevant environment, respectively. Suppose $hom(U/\eta, M)$ is an optimal internal variety of the system $\mathcal{S}$ with respect to $(\Phi, \mathcal{U})$. Then,*

1. *The optimal internal variety of the system $\mathcal{S}$ with respect to $(\Phi, \mathcal{U})$ is unique. That is, for an observation function $\eta' : U \to U/\eta$ if we have $\eta \neq \eta'$, then $hom(U/\eta', M)$ is not an optimal internal variety of the system $\mathcal{S}$ with respect to $(\Phi, \mathcal{U})$.*
2. *The optimal decision function in $hom(U/\eta', M)$ is unique.*

Proof Refer to Appendix 5.

2.5 Balance of External and Internal Variety

In order to investigate balance of external and internal variety, we need a couple of binary relations to compare the varieties. We begin with comparison of external varieties.

Let U and U' be subsets of $\hat{U}$. Let $\mathcal{U} = (U, \alpha)$ and $\mathcal{U}' = (U', \alpha')$ be environments and $\mathcal{S} = (\Phi, P, G, M, Y, U, \eta, d)$ and $\mathcal{S}' = (\Phi, P', G', M', Y', U', \eta', d')$ be goal seeking systems facing $\mathcal{U}$ and $\mathcal{U}'$, respectively. Suppose $E(\mathcal{S}, \mathcal{U}) \subset U \times U$ and $E(\mathcal{S}', \mathcal{U}') \subset U' \times U'$ are the external varieties of U and U' with respect to $\mathcal{S}$ and $\mathcal{S}'$, respectively.

Definition 7 The external variety $E(\mathcal{S}, \mathcal{U})$ of the environment $\mathcal{U}$ with respect to $\mathcal{S}$ is said to be greater than the external variety $E(\mathcal{S}', \mathcal{U}')$ of the environment $\mathcal{U}'$ with respect to $\mathcal{S}'$ if the followings hold:

1. $U' \subset U$.
2. $(\forall u \in U - U')([u] \subset U - U')$ and $(\forall u \in U')([u] \subset U')$, where $[u]$ denotes the equivalence class of u defined by the equivalence relation $E(\mathcal{S}, \mathcal{U})$.
3. $E(\mathcal{S}, \mathcal{U}) \cap (U' \times U') \subset E(\mathcal{S}', \mathcal{U}')$.

In this case we will write $E(\mathcal{S}, \mathcal{U}) \gg E(\mathcal{S}', \mathcal{U}')$.

The condition *1.* of the definition implies the size of the environment that the system has to process is increased from $\mathcal{U}'$ to $\mathcal{U}$. Usually only the present condition has been adopted to compare the varieties of the environments. The conditions *2.* says that $U - U'$, the increased part of the environment, is essentially different from U' in the sense of $E(\mathcal{S}, \mathcal{U})$. The condition *3.* implies that if u and u' in U' are similar in the sense of $E(\mathcal{S}, \mathcal{U})$, then so are they in the sense of $E(\mathcal{S}', \mathcal{U}')$. Hence, the conditions of the definition as a whole claim that if $\mathcal{U}$ is more "complex" than $\mathcal{U}'$, each state of U' is expected to be more "specific" in the sense of $E(\mathcal{S}, \mathcal{U})$ than of $E(\mathcal{S}', \mathcal{U}')$.

If we compare the external varieties given in Examples 1 and 2, then $\Delta \gg U \times U$ holds.

Definition 5 claims that for two environments $\mathcal{U} = (U, \alpha)$ and $\mathcal{U}' = (U', \alpha')$ even if the underlying sets U and U' are the same their varieties may be different. Due to this fact we would claim this definition can capture the intuitive meaning of the external variety of environment in a deeper sense than simply defining it just by the cardinality of the underlying set.

The following lemma will illustrate this situation.

Lemma 8 *Suppose an environment of a goal seeking system* $\mathcal{S} = (\Phi, P, G, M, Y, U, \eta, d)$ *changes from* $\mathcal{U} = (U, \alpha)$ *to* $\mathcal{U}' = (U, \alpha')$. *If*

$$(\forall t \in \{0, 1, 2, \dots\})(\exists t' \in \{0, 1, 2, \dots\})(\forall u \in U)(\alpha'^{t}(u) = \alpha^{t'}(u))$$

holds, then we have

$$E(\mathcal{S}, \mathcal{U}) \gg E(\mathcal{S}', \mathcal{U}').$$

Proof Refer to Appendix 6.

The assumption of this lemma intuitively represents that $\mathcal{U} = (U, \alpha)$ has more variety than $\mathcal{U}' = (U, \alpha')$ because any state of environment generated by α' can be realized by α even if the underlying set is the same. The condition certainly captures an aspect of intuition that $\mathcal{U} = (U, \alpha)$ is more complex than $\mathcal{U}' = (U, \alpha')$.

Now we turn to comparison of the internal varieties of systems.

Definition 8 Let U and U' be subsets of $\hat{U}$. Let $\mathcal{U} = (U, \alpha)$ and $\mathcal{U}' = (U', \alpha')$ be environments and $\mathcal{S} = (\Phi, P, G, M, Y, U, \eta, d)$ and $\mathcal{S}' = (\Phi, P', G', M', Y', U', \eta', d')$ be goal seeking systems facing $\mathcal{U}$ and $\mathcal{U}'$, respectively. The internal variety $hom(U/\eta, M)$ of $\mathcal{S}$ with respect to $(\Phi, \mathcal{U})$ is said to be greater than the internal variety $hom(U'/\eta', M')$ of $\mathcal{S}'$ with respect to $(\Phi, \mathcal{U}')$ if

1. $U' \subset U$
2. $|hom(U/\eta, M)| \geq |hom(U'/\eta', M')|$

hold, where $|\bullet|$ denotes the cardinality. In this case we will write $hom(U/\eta, M) \succeq hom(U'/\eta', M')$.

The condition *1.* refers to the size of the underlying set of the environment the system has to process, while the condition 2. is concerned with the size of the set of decision functions.

Especially if $U = U'$ and $M = M'$ hold, we have the following proposition concerning the fineness of η and the internal variety.

Proposition 9 *Let* U *and* U' *be subsets of* $\hat{U}$. *Let* $\mathcal{U} = (U, \alpha)$ *and* $\mathcal{U}' = (U', \alpha')$ *be environments and* $\mathcal{S} = (\Phi, P, G, M, Y, U, \eta, d)$ *and* $\mathcal{S}' = (\Phi, P', G', M', Y', U', \eta', d')$ *be goal seeking systems facing* $\mathcal{U}$ *and* $\mathcal{U}'$, *respectively. Suppose* $U = U'$ *and* $M = M'$*. Then, if* $\eta \subset \eta'$ *holds, then we have*

$$hom(U/\eta, M) \succeq hom(U/\eta', M).$$

Proof Refer to Appendix 7.

The proposition claims that a finer observation generates a larger internal variety.

Now we have a main result of this research, which represents our version of the phrase "only internal variety can destroy external variety."

Theorem 10 *Let* U *and* U' *be subsets of* $\hat{U}$*. Let* $\mathcal{U} = (U, \alpha)$ *and* $\mathcal{U}' = (U', \alpha')$ *be environments and* $\mathcal{S} = (\Phi, P, G, M, Y, U, \eta, d)$ *and* $\mathcal{S}' = (\Phi, P', G', M', Y', U', \eta', d')$ *be goal seeking systems facing* $\mathcal{U}$ *and* $\mathcal{U}'$*, respectively.*

Suppose $\mathcal{S}$ *and* $\mathcal{S}'$ *are irredundant with respect to* $(\Phi, \mathcal{U})$ *and* $(\Phi, \mathcal{U}')$*, respectively, and* $U' \subset U$ *and* $\phi'_{\mathcal{S}} = \phi_{\mathcal{S}}|U'$*. Assume* $hom(U/\eta, M)$ *and* $hom(U'/\eta', M')$ *are optimal internal varieties of* $\mathcal{S}$ *and* $\mathcal{S}'$ *with respect to* $(\Phi, \mathcal{U})$ *and* $(\Phi, \mathcal{U}')$*, respectively. Then if*

$$E(\mathcal{S}, \mathcal{U}) \gg E(\mathcal{S}', \mathcal{U}')$$

holds, we have

$$hom(U/\eta, M) \succeq hom(U'/\eta', M').$$

Proof Refer to Appendix 8.

Now we examine the meaning of Theorem 10. Suppose a goal seeking system $\mathcal{S}'$ possesses an optimal internal variety $E(\mathcal{S}', \mathcal{U}')$ with respect to $(\Phi, \mathcal{U}')$. Suppose $\mathcal{U}'$ changes to $\mathcal{U}$ and thus the variety of the environment changes from $E(\mathcal{S}', \mathcal{U}')$ to $E(\mathcal{S}, \mathcal{U})$. Then, in order to attain an optimal internal variety the system has to increase the variety from $hom(U'/\eta', M')$ to $hom(U/\eta, M)$ as far as the system is irredundant.

Furthermore, if the system cannot increase its internal variety, then the system has to maintain the variety of the environment as $E(\mathcal{S}', \mathcal{U}')$.

If it is impossible and the variety increases from $E(\mathcal{S}', \mathcal{U}')$ to $E(\mathcal{S}, \mathcal{U})$ C$hom(U'/\eta', M')$ fails to be optimal with respect to $(\Phi, \mathcal{U})$.

2.6 Conclusion

In this article we generalize and formulate the Law of Requisite Variety by formalizing the key concepts, i.e., external and internal variety, in the context of relationship between a goal seeking system and its environment in the framework of logico-mathematical general systems theory.

The generalization is twofold: First, we treat dynamic state transition of the environment. So far "variety" of a system has been basically understood as the number of the distinguishable possible states of it so that there have been no explicit discussions on trajectory of states of the environment.

Second, we introduce idea of subjective variety. In the discussions so far variety is basically "objectively" defined as the number of possible states. Our formulation emphasizes that the external (environmental) variety should not be only determined by the characteristics of the environment itself but also be affected by the "value system" of the system. On the other hand, the internal variety of the system is formalized in such a way that if its observation ability and/or the size of the set of alternatives increase, then so does the internal variety of the system, as far as the underlying set of the environment is fixed.

By using the concepts of these varieties represented in terms of equivalence relations, we explored the relationship between a goal seeking system and its relevant environment to derive a mathematical version of "only the internal variety can destroy the external variety" in a rigid way.

We believe the formalizations and discussions developed here satisfy our intuition and provide deeper insights, and the results go beyond it.

Appendix

1. Proof of Proposition 1

We will first define $\beta : M \to M$ and then show that β satisfies the condition.

Let $m \in M$ be arbitrary. If there is $u \in U$ such that $\phi(u) = m$, we define β by$\beta(m) = \phi\alpha(u)$. If there is no such u in U, then let $\beta(u)$ be arbitrary in M. Then, β is well-defined: Suppose u and u' are in U such that $\phi(u) = m = \phi(u')$. Due to System–environment consistency assumption, we have $\phi\alpha(u) = \phi\alpha(u')$ and hence β is well-defined.

Next, we prove β satisfies the desired condition by induction on t. Due to initial solvability we have $\phi(u) = d\eta(u)$ for every $u \in U$. Furthermore, β clearly satisfies

$$(\forall u \in U)(\phi\alpha(u) = \beta(m) = \beta\phi(u)).$$

Hence, the cases of $t = 0$ and $t = 1$ obviously hold.

Now assume the statement holds for $t = i - 1$. Let $u \in U$ be arbitrary. Then we have

$$\beta^i\phi(u) = \beta(\beta^{i-1}\phi(u)) = \beta\phi\alpha^{i-1}(u).$$

By the definition of β it follows that

$$\beta\phi\alpha^{i-1}(u) = \phi\alpha\alpha^{i-1}(u) = \phi\alpha^{i}(u),$$

which completes the proof.

□

2. *Proof of Corollary 2*
Let us suppose that β and β' are internal models of $\mathcal{U}$. Then, we have $\beta^t\phi = \phi\alpha^t = \beta'^t\phi$ for every $t \in T$. Since ϕ is surjective it follows that $\beta^t = \beta'^t$ for every $t \in T$. By letting $t = 1$ we have $\beta = \beta'$.

□

3. *Proof of Proposition 3*
Suppose the assumptions are satisfied and β is an internal model of $\mathcal{U}$ with respect to Φ. From the assumption we have

$$(\forall u \in U)(\forall t \in \{0, 1, 2, \dots\})(\beta^t d\eta(u) = \phi\alpha^t(u)).$$

By setting $t = 0$ we have $d\eta(u) = \phi(u)$ holds for every $u \in U$ and hence the system has initial solvability.

Now it only remains to show $\phi \subset \phi\alpha$. Let $(u, u') \in \phi$. Then it follows from $\phi(u) = \phi(u')$ that

$$\begin{aligned}\phi\alpha(u) &= \beta d\eta(u) = \beta\phi(u) = \beta\phi(u') \\ &= \beta d\eta(u') = \phi\alpha(u')\end{aligned}$$

since $d\eta(u) = \phi(u)$ holds for every $u \in U$ due to initial solvability. Consequently, we have $\phi \subset \phi\alpha$ and hence the system and the environment are consistent.

□

4. *Proof of Proposition 6*

If Part Let $\eta = E(\mathcal{S}, \mathcal{U})$. We will show that there is an optimal decision function with respect to $(\Phi, \mathcal{U})$ in $hom(U/E(\mathcal{S}, \mathcal{U}), M)$.

Let us define $d : U/E(\mathcal{S}, \mathcal{U}) \to M$ by

$$d([u]) = \phi(u) \text{ for every } [u] \in U/E(\mathcal{S}, \mathcal{U}).$$

We will verify that d is an optimal decision function with respect to $(\Phi, \mathcal{U})$. Indeed, let $(u, u') \in E(\mathcal{S}, \mathcal{U})$, then by definition

$$(\forall t \in \{0, 1, 2, \dots\})(\phi\alpha^t(u) = \phi\alpha^t(u'))$$

holds, which implies $\phi(u) = \phi(u')$ and hence d is well-defined. Furthermore, (η, d) clearly has initial solvability since

$$d\eta(u) = d([u]) = \phi(u) \text{ holds for every } u \in U.$$

Suppose (η', d') is such that $d'\eta'(u) = \phi(u)$ holds for every $u \in U$, where $\eta' : U \to U/\eta'$ and $d' : U/\eta' \to M$. We will show that there is a natural map $\rho : U/\eta' \to U/\eta$ such that Fig. 2.5 is commutative.

Let $(u, u') \in \eta'$. Since (η', d') has initial solvability Cit follows from Proposition 1 that there is $\beta : M \to M$ such that

$$(\forall t \in \{0, 1, 2, \dots\})(\beta^t d'\eta'(u) = \phi\alpha^t(u)).$$

On the other hand, $\eta'(u) = \eta'(u')$ implies

$$(\forall t \in \{0, 1, 2, \dots\})(\beta^t d'\eta'(u) = \beta^t d'\eta'(u')).$$

Thus, $\phi\alpha^t(u) = \beta^t d'\eta'(u) = \beta^t d'\eta'(u') = \phi\alpha^t(u')$ holds for every t and hence $(u, u') \in E(\mathcal{S}, \mathcal{U})$. Consequently, we have $\eta' \subset \eta$ and hence we can define a natural map $\rho : U/\eta' \to U/\eta$ by $\rho([u]') = [u]$, where $[u]$ and $[u]'$ denote the equivalence classes of u with respect to η and η', respectively.

Finally, we will show ρ makes Fig. 2.5 commutative. Since $(\forall u \in U)(\rho\eta'(u) = \rho([u]') = [u]$ holds and (η, d) and (η', d') have initial solvability,

$$(\forall u \in U)(d\rho\eta'(u) = d\eta(u) = \phi(u) = d'\eta'(u))$$

holds and hence we have $d\rho\eta' = d'\eta'$. Since η' is surjective, we have $d\rho = d'$.

Only If Part Suppose $hom(U/\eta, M)$ is an optimal internal variety of the system with respect to $(\Phi, \mathcal{U})$. By definition there is an optimal decision function $d : U/\eta \to M$ in $hom(U/\eta, M)$.

To show $\eta = E(\mathcal{S}, \mathcal{U}))$ we will verify $\eta \subset E(\mathcal{S}, \mathcal{U})$ first. Suppose $(u, u') \in \eta$ holds. Since (η, d) has initial solvability, by Proposition 1 there is $\beta : M \to M$ such that

$$(\forall t \in \{0, 1, 2, \dots\})(\beta^t d\eta(u) = \phi\alpha^t d\eta(u))$$

holds for every $u \in U$. Consequently, we have

$$\begin{aligned} & (u, u') \in \eta \\ \to\ & \eta(u) = \eta(u') \\ \to\ & (\forall t \in \{0, 1, 2, \dots\})(\beta^t d\eta(u) = \beta^t d\eta(u')) \end{aligned}$$

$$\rightarrow (\forall t \in \{0, 1, 2, \ldots\})(\phi\alpha^t(u) = \phi\alpha^t(u'))$$
$$\rightarrow (u, u') \in E(\mathcal{S}, \mathcal{U}).$$

Next, conversely, we show $E(\mathcal{S}, \mathcal{U}) \subset \eta$. Suppose that $(u, u') \in E(\mathcal{S}, \mathcal{U})$ holds. Then, by definition

$$(\forall t \in \{0, 1, 2, \ldots\})(\phi\alpha^t(u) = \phi\alpha^t(u')).$$

By setting $t = 0$ we have $\phi(u) = \phi(u')$. Now since $(\forall u \in U)(1\phi(u) = \phi(u))$ trivially holds, where 1 denotes an identity on M, $(\phi, 1)$ has initial solvability. Since d is an optimal decision function, by the condition 2. of Definition 5 there is $\rho : M \rightarrow U/\eta$ such that

$$(\forall u \in U)(\rho\phi(u) = \eta(u))$$

holds. Consequently, it follows from $\phi(u) = \phi(u')$ that

$$\eta(u) = \rho\phi(u) = \rho\phi(u') = \eta(u'),$$

which implies $(u, u') \in \eta$.

□

5. Proof of Lemma 7

1. Obvious from Proposition 6.
2. Let d and $d' \in hom(U/\eta, M)$ be optimal decision functions with respect to $(\Phi, \mathcal{U})$. Since $hom(U/\eta, M)$ is optimal, (η, d) and (η, d') have initial solvability. By Proposition 1 there are $\beta : M \rightarrow M$ and $\beta' : M \rightarrow M$ such that

$$(\forall u \in U)(\forall t \in \{0, 1, 2, \ldots\})(\beta^t d\eta(u) = \phi\alpha^t(u'))$$

and

$$(\forall u \in U)(\forall t \in \{0, 1, 2, \ldots\})(\beta'^t d'\eta(u) = \phi\alpha^t(u')).$$

By setting $t = 0$ it follows that $d\eta = \phi = d'\eta$. Since η is surjective, it implies $d = d'$.

□

6. Proof of Lemma 8

Since the conditions 1. and 2. of Definition 5 clearly hold it is sufficient to show $E(\mathcal{S}, \mathcal{U}) \subset E(\mathcal{S}, \mathcal{U}')$.

Let $(u, u') \in E(\mathcal{S}, \mathcal{U})$, then by definition we have

$$(\forall t \in \{0, 1, 2, \ldots\})(\phi\alpha^t(u) = \phi\alpha^t(u')).$$

We will show $(u, u') \in E(\mathcal{S}, \mathcal{U}')$, that is,

$$(\forall t \in \{0, 1, 2, \dots\})(\phi\alpha''^t(u) = \phi\alpha''^t(u')).$$

However, for every $t \in \{0, 1, 2, \dots\}$ there is $t' \in \{0, 1, 2, \dots\}$ such that for any $u \in U$, $\alpha''^t(u) = \alpha^{t'}(u)$ holds. Consequently, we have

$$\phi\alpha''^t(u) = \phi\alpha^{t'}(u) = \phi\alpha^{t'}(u') = \phi\alpha''^t(u').$$

□

7. Proof of Proposition 9

Since the condition 1. of Definition 8 obviously holds, we are only required to show that $|hom(U/\eta', M)| \geq |hom(U/\eta, M)|$. We verify the statement by showing that there is an injection from $hom(U/\eta, M)$ to $hom(U'/\eta, M)$. Since $\eta \subset \eta'$ holds there is a natural map $\rho : U/\eta \to U/\eta'$. Now let us define $F : hom(U/\eta', M) \to hom(U/\eta, M)$ by $F(d') = d'\rho$ for every $d' \in hom(U/\eta', M)$. Then F is injective: Indeed, if $F(d') = F(d'')$ holds, then we have $d'\rho = d''\rho$, which implies $d' = d''$, since ρ is surjective.

□

8. Proof of Theorem 10

Let us assume $E(\mathcal{S}, \mathcal{U}) \gg E(\mathcal{S}', \mathcal{U}')$. By the assumption and Proposition 6 we have $\eta = E(\mathcal{S}, \mathcal{U})$ and $\eta' = E(\mathcal{S}', \mathcal{U}')$. We are only required to show that $|hom(U/\eta, M)| \geq |hom(U/\eta', M')|$ since $U' \subset U$ holds by the assumption. We will verify it by constructing an injection

$$F : hom(U'/\eta', M') \to hom(U/\eta, M).$$

Since $\phi_{\mathcal{S}}$ and $\phi_{\mathcal{S}'}$ are surjective, $\phi_{\mathcal{S}}(U) = M$ and $\phi_{\mathcal{S}'}(U') = M'$ hold. $U' \subset U$ implies $M' \subset M$ and hence there is an inclusion $i : M' \to M$.

Now let us define $\rho : U/\eta \to U'/\eta'$ by

$$\rho([u]) = \begin{cases} [u]' & (\text{if } u \in U') \\ [u^*]' & (\text{otherwise}) \end{cases}$$

where u^* is arbitrary in U' while $[u]$ and $[u]'$ denote the equivalence classes of u with respect to $\phi_{\mathcal{S}}$ and $\phi_{\mathcal{S}'}$, respectively.

Then the definition is proper: Indeed, let u and v be in U such that $[u] = [v]$. We can verify $\rho([u]) = \rho([v])$ as follows: Suppose $u \in U - U'$ then by definition we have $[u] \subset U - U'$, which implies $[v] \subset U - U'$ and hence $v \in U - U'$. Similarly, if $u \in U'$ holds, then we have $[u] \subset U'$, so that $v \in U'$. Consequently, we have either $u, v \in U - U'$ or $u, v \in U'$ if $[u] = [v]$ holds.

If u and v are in $U - U'$, then we have

$$\rho([u]) = [u^*]' = \rho([v]).$$

On the other hand, if u and v are in U' then we have $(u, v) \in E(\mathcal{S}', \mathcal{U}')$, i.e., $[u]' = [v]'$ holds, because of $[u] = [v]$ and the condition 3. of Definition 5, i.e.,

$$E(\mathcal{S}, \mathcal{U}) \cap (U' \times U') \subset E(\mathcal{S}', \mathcal{U}').$$

Hence, it follows that

$$\rho([u]) = [u]' = [v]' = \rho([v]).$$

Furthermore, ρ is surjective: Let $[u]' \in U'/\eta'$ be arbitrary. Then since $u \in U' \subset U$ holds, by taking $[u] \in U/\eta$, we have $\rho([u]) = [u]'$.

Now let us define F by

$$F(d') = id'\rho$$

for every $d' \in hom(U'/\eta', M')$. Then, F is clearly a function from $hom(U'/\eta', M')$ to $hom(U/\eta, M)$. Furthermore, F is injective: Suppose $F(d') = F(d'')$ then by definition we have $id'\rho = id''\rho$. Since i is injective and ρ is surjective, we have $d' = d''$.

□

References

Ashby, R. (1963). *An Introduction to Cybernetics*. John Wiley.

Beer, S. (1981). *Brain of the Firm* (2nd ed.). John Wiley. International Society for Systems Sciences (ISSS). 2021. https://www.isss.org/about-isss/

Kickert, W. J. M. (1980). *Organization of Decision Making—A Systems Theoretic Approach*. North-Holland.

Kijima, K. (1983). Formulation of the law of requisite variety (in Japanese). In *Proceeding of the Operations Research Society of Japan*.

Kijima, K. (1984). *Algebraic formulation of relationship between system and its environment—Based on law of requisite variety and internal model principle* (in Japanese). Technical Report J3, Department of Management Science and Engineering, Tokyo Institute of Technology.

Kijima, K., Takahara, Y., & Nakano, B. (1985). Algebraic formulation of internal model principle. In *Proceeding of the 29th Annual Meeting of SGSR*, Los Angeles.

Kijima, K., Takahara, Y., & Nakano, B. (1986). Algebraic formulation of relationship between a goal seeking system and its environment. *International Journal of General System*, *12*(4), 341–358.

Klir, G. (2001). *Facets of systems science* (2nd ed.). Kluwer Academic/Plenum Publishers.

Mesarovic, M. D. (1967). General system theory and its mathematical foundation. In *Systems Science and Cybernetics Conference, IEEE, Boston, Massachusetts*.

Miyazawa, K. (Ed.). (1983). Decision process and organization structure (in Japanese). In K. Miyazawa (Ed.), *Management decision making*. Diamond-sha.

Nakano, B. (1983). Relationship between system and its environment and internal modelling (in Japanese). In *Proceeding of Systems Symposium in 1983*.

Wonham, W. N. (1976). Towards an abstract internal model principle. *IEEE Transactions on Systems, Man, and Cybernetics*, *IEEE-SMC* 6(11), 735–740.

Chapter 3
Isomorphy of Subsystem and Component Subsystem of Input–Output System

Junichi Iijima

Abstract Although the concept of subsystems and that of component subsystems are quite different, both viewpoints to a part of a system can be identified in some case. In such a case, we can handle a subsystem and a component subsystem, interchangeably. In this paper, we will focus on the condition for an input–output system under which we can identify both of them. Firstly, we review mathematical foundations for the discussion. Secondly, we will investigate algebraic properties for which a subalgebra and a quotient algebra can be identified. Finally, we adopt the discussion to input–output systems with a particular algebraic structure.

Keywords Algebra · Normal subalgebra · Subsystem · Component subsystem · Isomorphy

3.1 Introduction

A subsystem is often defined as a subset of a given system, while a component subsystem is one of the systems of which a complex system consists of. In this sense, there is a crucial difference between the concept of subsystems and that of component subsystems. Wymore (1976) has also pointed out that it is of great importance to note the difference.

It can be said that systems theory is originally proposed in the complementary side of the reductionism. A well-known statement in systems thinking is

This essay is an extended version of the paper "Equivalence of Subsystems and Components" in *Int. J. General Systems*, Vol. 24(3), PP. 271–279, 1996 by Junichi Iijima.

J. Iijima (✉)
Department of International Digital and Design Management, School of Management, Tokyo University of Science, Tokyo, Japan
e-mail: junichi.iijima@rs.tus.ac.jp

K. Kijima et al. (eds.), *Systems Research I*, Translational Systems Sciences 26,
https://doi.org/10.1007/978-981-16-9937-5_3

"The Whole is greater than the sum of its parts" by Aristotle. Symbolically, it can be described as

$$S \subset S_1 \cup \cdots \cup S_n$$

Reductionism approach has got a great success in natural science, especially in Physics. It is because several aspects of phenomena can be investigated separately. For example, "... we separate out the phenomena of heat from the total complexity of the physical world..." (Checkland, 1981). Symbolically, it can be described as

$$S \subset S_1 \times \cdots \times S_n$$

In the above sense, an assumption in reductionist approach can be stated that a system as a whole can be described as the sum of subsystems and/or the product of component subsystems.

As we will see, those two viewpoints for "sub" concepts can be identified in some case. In such a case, we can handle a subsystem and a component subsystem, interchangeably. In this essay, we will focus on the conditions for input–output systems under which we can identify both of them.

Since the discussion developed in this essay is based on algebraic concepts, we firstly, review algebraic concepts for discussion. The concept of normal subgroups gives us a crucial hint to solve the problem. Secondly, we will investigate the condition under which a subalgebra and a quotient algebra can be identified. Finally, we adopt the discussion to input–output systems with a particular algebraic structure.

3.2 Mathematical Preliminaries

Let us firstly review several mathematical concepts of algebras that are important to develop our consideration.

Definition 2.1 (Algebra) An **algebra** is a couple $\mathscr{A} =< A; F >$, where A is a nonempty set called an underlying set and F is a class of finitary operations on A. From now on, we denote an algebra by script font $\mathscr{A}$ or $\mathscr{B}$ and we will index F by $F = \{f_\gamma \mid \gamma \in \Gamma\}$.

Definition 2.2 (Type) The **type** of an algebra $\mathscr{A} =< A; F >$ is a sequence of natural numbers $< n_0, \ldots, n_\gamma, \ldots >$ such that each $f_\gamma \in F$ is an n_γ-ary operation on A.

The most important methods for constructing a new algebra from a given class of algebras are as follows:

- subalgebras
- homomorphic images
- products

Let us next review these concepts.

Definition 2.3 (Subalgebra) Let $\mathscr{A} =< A; F >$ be an algebra and $B \subset A$. If B is closed under all of f_γ in Γ, then an algebra$< B; F_B >$ is called a **subalgebra** of $\mathscr{A}$, where $F_B = \{f_\gamma|_B \mid f_\gamma \in F\}$. We will denote f_γ instead of $f_\gamma|_B$ and describe F instead of F_B if there is no confusion.

Let us next review the concept of homomorphic images.

Definition 2.4 (Homomorphism) Let $\mathscr{A} =< A; F >$ and $\mathscr{B} =< B; F >$ be algebras of the same type. And let $h : A \to B$ be a mapping of A into B. Then h is called a **homomorphism** of $\mathscr{A}$ into $\mathscr{B}$ if and only if the following condition holds:

$$
\begin{aligned}
&(\forall f_\gamma \in F)(\forall a_0, \ldots, a_{n_\gamma - 1} \in A)(f_\gamma(h(a_0), \ldots, h(a_{n_\gamma - 1})) \\
&\quad = h(f_\gamma(a_0, \ldots, a_{n_\gamma - 1})))
\end{aligned}
$$

In this case, $< h(A); F >$ is a subalgebra of $\mathscr{B}$ and we call it as a **homomorphic image** of $\mathscr{A}$ under h. If a homomorphism h of $\mathscr{A}$ into $\mathscr{B}$ is bijective, it is called an **isomorphism** and $< h(A); F >= \mathscr{B}$ is called an **isomorphic image** of $\mathscr{A}$ under h. If there exists an isomorphism from $\mathscr{A} =< A; F >$ onto $\mathscr{B} =< B; F >$, we denote it by $\mathscr{A} \cong \mathscr{B}$.

Then let us next review the concept of products.

Definition 2.5 (Product) Let $\boldsymbol{A} = \{\mathscr{A}_i | \mathscr{A}_i =< A_i; F >, i \in I\}$ be a class of algebras of the same type. Then the **product** of $\boldsymbol{A}$ is an algebra $\boldsymbol{\Pi A} =< \Pi(A_i \mid i \in I); F >$. And a subalgebra $\mathscr{B} =< B; F >$ of the product $\boldsymbol{\Pi A}$ satisfying $p_i(B) = A_i$ for all $i \in I$, where $p_i : B \to A_i$ is the i-th projection, is called a **subdirect product** of $\boldsymbol{A}$. In this case, we call $\mathscr{A}_i$ as a component subalgebra of $\mathscr{B}$.

One more important construction is a quotient algebra induced by a congruence relation.

Definition 2.6 (Congruence Relation) Let $\mathscr{A} =< A; F >$ be an algebra and $\Theta \subset A \times A$ an equivalence relation on A, that is, a binary relation on A satisfying the following conditions:

1. Reflexivity

$$(\forall x \in A)(x \Theta x)$$

2. Symmetry

$$(\forall x, y \in A)(x \Theta y \Rightarrow y \Theta x)$$

3. Transitivity

$$(\forall x, y, z \in A)(x \Theta y \wedge y \Theta z \Rightarrow x \Theta z)$$

If Θ satisfies the following condition called the substitution property, then it is called a **congruence relation** on $\mathscr{A}$:

$$\text{for any } f_\gamma \in F, a_0, \ldots, a_{n_\gamma-1}, b_0, \ldots, b_{n_\gamma-1} \in A,$$

$$a_0 \Theta b_0 \wedge \ldots \wedge a_{n_\gamma-1} \Theta b_{n_\gamma-1} \Rightarrow f_\gamma(a_0, \ldots, a_{n_\gamma-1}) \Theta f_\gamma(b_0, \ldots, b_{n_\gamma-1})$$

We can construct a new algebra with a congruence relation.

Definition 2.7 (Quotient Algebra) Let $\mathscr{A} =< A; F >$ be an algebra and $\Theta \subset A \times A$ a congruence relation on $\mathscr{A}$. Let A/Θ be the quotient set of A by Θ, where $A/\Theta = \{[a]_\Theta \mid a \in A\}$. Then $< A/\Theta; F >$ is an algebra of the same type as $\mathscr{A}$, where $A/\Theta = \{[a]_\Theta \mid a \in A\}$ and $f_\gamma \in F$ is defined by $f_\gamma([a_0]_\Theta, \ldots, [a_{n_\gamma-1}]_\Theta) = [f_\gamma(a_0, \ldots, a_{n_\gamma-1})]_\Theta$ for any $a_0, \ldots, a_{n_\gamma-1} \in A$. We call it as the **quotient algebra** of $\mathscr{A}$ by Θ and denote it by $\mathscr{A}/\Theta$.

From now on, we will simply describe each equivalence class $[a]_\Theta$ by $[a]$.

Suppose that $\mathscr{B} =< B; F >$ is a subdirect product of $\mathbf{A}$, where $\mathbf{A} = \{\mathscr{A}_i | \mathscr{A}_i =< A_i; F >, i \in I\}$ is a class of algebras of the same type. Since $p_i : B \rightarrow A_i$ is a homomorphism of $\mathscr{B} =< B; F >$ onto $\mathscr{A}_i =< A_i; F >$, $Ker(p_i) \subset B \times B$ is a congruence relation on $\mathscr{B} =< B; F >$, where $Ker(p_i)$ is defined by $Ker(p_i) = \{(b, b') \mid (p_i(b) = p_i(b')\}$. Since $\mathscr{A}_i \cong \mathscr{B}/Ker(p_i)$, we can consider a quotient algebra as a component subalgebra of the subdirect product.

Since the concept of normal subgroups gives a crucial hint to solve the problem mentioned earlier, let us next survey several concepts on groups.

Definition 2.8 (Semigroup) A **semigroup** is an algebra $< A, \{\cdot\} >$ with a binary operation "$\cdot$" on A satisfying the following condition called associativity:

$$(\forall x, y, z \in A)((x \cdot y) \cdot z = x \cdot (y \cdot z))$$

If an algebra $\mathscr{A} =< A; F >$ has a binary operation "@" and associative on the operation, we may call $\mathscr{A}$ a **semigroup on "@"**. From now on, we focus on semigroups in this sense and explicitly mention "@" as $\mathscr{A} =< A; \{@\} \cup F >$.

Definition 2.9 (Group) A **group** is an algebra $\mathscr{G} =< G, \{\cdot, ^{-1}, e\} >$, where "$\cdot$" is a binary operation on G, "$^{-1}$" is a unary operation on G, and "e" is a nullary operation on G such that the following conditions are satisfied:

1. Semigroup on "$\cdot$"
2. Identity

$$(\forall x \in G)(e \cdot x = x \cdot e = x)$$

3. Inverse

$$(\forall x \in G)(x \cdot x^{-1} = x^{-1} \cdot x = e)$$

A subgroup of a given group is a subalgebra of the group in the sense of Definition 2.3. Among subgroups, the concept of normal subgroups plays a crucial role in this essay.

Definition 2.10 (Normal Subgroup) Let $\mathcal{G} =< G; \{\cdot, ^{-1}, e\} >$ be a group and $\mathcal{N} =< N; \{\cdot, ^{-1}, e\} >$ a subgroup of $\mathcal{G}$. Then $\mathcal{N}$ is called a **normal subgroup** of $\mathcal{G}$ if it satisfies the following condition:

$$(\forall n \in N)(\forall g \in G)(gng^{-1} \in N)$$

One of the important properties of normal subgroups is that we can construct a congruence relation based on it.

Proposition 2.1 *Let $\mathcal{G} =< G; \{\cdot, ^{-1}, e\} >$ be a group and $\mathcal{N} =< N; \{\cdot, ^{-1}, e\} >$ a normal subgroup of $\mathcal{G}$. Let us define a binary relation $\sim_{\mathcal{N}}$ on G by*

$$x \sim_{\mathcal{N}} x' \Leftrightarrow (\exists g \in G)(x, x' \in Ng)$$

Then $\sim_{\mathcal{N}}$ is a congruence relation on $\mathcal{G}$.

All of the propositions, corollaries, and theorems in this section are well-known, and therefore, we do not state the proofs of them. Standard textbooks on algebras (MacLane and Birkoff, 1974) and universal algebras (Graetzer, 1968) should be consulted.

Corollary 2.1 *Let $\mathcal{G} =< G; \{\cdot, ^{-1}, e\} >$ be a group and $\mathcal{N} =< N; \{\cdot, ^{-1}, e\} >$ a normal subgroup of $\mathcal{G}$. Then $\mathcal{G}/\sim_{\mathcal{N}}$ is a group. We often denote $\mathcal{G}/\sim_{\mathcal{N}}$ by $\mathcal{G}/\mathcal{N}$.*

By the above corollary, we have a quotient group induced by a normal subgroup. Since a quotient group can be considered as a "component" group in a subdirect factorization of a given group, we can identify a subgroup with a "component" if the subgroup is isomorphic to some quotient group. Then the following theorem states the condition under which it is the case (MacLane and Birkoff, 1974).

Theorem 2.1 *Let $\mathcal{G} =< G; \{\cdot, ^{-1}, e\} >$ be a group and $\mathcal{H} =< H; \{\cdot, ^{-1}, e\} >$ a subgroup of $\mathcal{G}$. If there is a normal subgroup $\mathcal{N} =< N; \{\cdot, ^{-1}, e\} >$ of $\mathcal{G}$ such that $N \cap H = \{e\}$ and $G \subset \{n \cdot h \mid n \in N \wedge h \in H\}$, then $\mathcal{G}/\mathcal{N} \cong \mathcal{H}$ holds.*

In this essay, our objective is to find the condition for input–output systems under which we can treat a subsystem and a component system, interchangeably. The above theorem provides a good model for our investigation.

3.3 Normal Subalgebras

Let us next investigate under what conditions we can identify a subalgebra as a quotient algebra.

Definition 3.1 (Normal Subalgebra) Let $\mathscr{A} =< A; \{@\} \cap F >$ be a semigroup on "@", that is, "@" is a binary operation and $\mathscr{A}$ is associative on "@". Then a subalgebra $\mathscr{B} =< B; \{@\} \cap F >$ of $\mathscr{A}$ is called a **normal subalgebra** of $\mathscr{A}$ with respect to "@" if it satisfies the following conditions:

1. $(\forall f_\gamma \in \{@\} \cup \in F)(\forall a_0, \ldots, a_{n_\gamma - 1} \in A)(\forall b_0, \ldots, b_{n_\gamma - 1} \in B)(\exists b \in B)$
   $(f_\gamma(a_0 @ b_0, \ldots, a_{n_\gamma - 1} @ b_{n_\gamma - 1}) = f_\gamma(a_0, \ldots, a_{n_\gamma - 1}) @ b)$
2. $(\forall a, a' \in A)(\forall b, b' \in B)(\exists b'' \in B)(a @ b = a' @ b' \Rightarrow a = a' @ b'')$

There are several examples of normal subalgebras.

Example 3.1 (Normal Subgroup) A normal subgroup of a group is a normal subalgebra. Actually, let $\mathscr{G} =< G; \{\cdot, ^{-1}, e\} >$ be a group and $\mathscr{N} =< N; \{\cdot, ^{-1}, e\} >$ a normal subgroup of $\mathscr{G}$. Suppose that "·" is the focal binary operation @ and in this case, $F = \{^{-1}, e\}$. Then

1. For any $a_0, a_1 \in G, b_0, b_1 \in N$, there exists $b_2, b \in N$ such that $(a_0 \cdot b_0) \cdot (a_1 \cdot b_1) = a_0 \cdot (b_0 \cdot a_1) \cdot b_1 = a_0 \cdot (a_1 \cdot b_2) \cdot b_1 = (a_0 \cdot a_1) \cdot (b_2 \cdot b_1) = (a_0 \cdot a_1) \cdot b$ for $\mathscr{N}$ is a normal subgroup.
2. Since e is a nullary operation, the condition obviously holds.
3. For any $a_0 \in G, b_0 \in N$, there exists $b \in N$ such that $(a_0 \cdot b_0)^{-1} = b_0^{-1} \cdot a_0^{-1} = a_0^{-1} \cdot b$ for $\mathscr{N}$ is a normal subgroup.

Therefore the first condition holds. It is obvious that the second condition is also satisfied for $\mathscr{G}$ is a group and $\mathscr{N}$ is a normal subgroup. Consequently, $\mathscr{N}$ is a normal subalgebra of $\mathscr{G}$ with respect to "·".

Example 3.2 (Ideal of Ring) An ideal of a ring is a normal subalgebra. Actually, let $\mathscr{R} =< R; \{+, -, 0, \cdot\} >$ be a ring and $\mathscr{I} =< I; \{+, -, 0, \cdot\} >$ an ideal of $\mathscr{R}$. Suppose that "+" is the focal binary operation "@", in this case, $F = \{-, 0, \cdot\}$. Then

1. For any $a_0, a_1 \in R, b_0, b_1 \in I$, there exists $b \in I$ such that $(a_0+b_0)+(a_1+b_1) = a_0+(b_0+a_1)+b_1 = a_0+(a_1+b_0)+b_1 = (a_0+a_1)+(b_0+b_1) = (a_0+a_1)+b$ holds where $b = (b_0 + b_1) \in I$ for "+" is associative and commutative and $\mathscr{I}$ is an ideal of $\mathscr{R}$.
2. Since 0 is a nullary operation, the condition obviously holds.
3. For any $a_0 \in R, b_0 \in I, -(a_0 + b_0) = -a_0 + (-b_0) = (-a_0) + b$ holds, where $b = -b_0 \in I$.
4. For any $a_0, a_1 \in R, b_0, b_1 \in I, (a_0+b_0) \cdot (a_1+b_1) = a_0 \cdot a_1 + a_0 \cdot b_1 + b_0 \cdot a_1 + b_0 \cdot b_1$. Since I is an ideal, $a_0 \cdot b_1, b_0 \cdot a_1, b_0 \cdot b_1 \in I$. Therefore there exists $b = a_0 \cdot b_1 + b_0 \cdot a_1 + b_0 \cdot b_1 \in I$ such that $(a_0 + b_0) \cdot (a_1 + b_1) = (a_0 \cdot a_1) + b$.

Consequently, the first condition holds. It is obvious that the second condition is also satisfied for $\mathscr{R}$ is a ring and $\mathscr{I}$ is an idea. Hence, $\mathscr{I}$ is a normal subalgebra of $\mathscr{R}$ with respect to "+".

Example 3.3 (Linear Subspace) A linear subspace of a linear space is a normal subalgebra. Actually, let $\mathscr{V} =< V; \{+, 0, -, \cdot, 1\} \cup \Gamma >$ be a linear space and $\mathscr{W} =< W; \{+, 0, -, \cdot, 1\} \cup \Gamma >$ a linear subspace of $\mathscr{V}$, where Γ is the set of scalar multiplication. Suppose that "+" is the focal binary operation "@" and in this case, $F = \{0, -, \cdot, 1\} \cup \Gamma$. Since $\mathscr{W}$ is a normal subgroup of $\mathscr{V}$, it can be easily shown by Example 3.1.

Example 3.4 (Some Ideals of Distributive Lattice) A special type of ideals of a distributive lattice is a normal subalgebra. Actually, let $\mathscr{L} =< L; \{\wedge, \vee\} >$ be a distributive lattice and $\mathscr{I} =< I; \{\wedge, \vee\} >$ an ideal of $\mathscr{L}$. Suppose that "$\vee$" is the focal binary operation "@", in this case $F = \{\wedge\}$. If the second condition holds, then $\mathscr{I}$ is a normal subalgebra of $\mathscr{L}$. Because

1. For any $a_0, a_1 \in L, b_0, b_1 \in I$, there exists $b \in I$ such that $(a_0 \vee b_0) \vee (a_1 \vee b_1) = a_0 \vee (b_0 \vee a_1) \vee b_1 = a_0 \vee (a_1 \vee b_0) \vee b_1 = (a_0 \vee a_1) \vee (b_0 \vee b_1) = (a_0 \vee a_1) \vee b$ holds, where $b = (b_0 \vee b_1) \in I$ for "$\vee$" is associative and commutative, and $\mathscr{I}$ is an ideal of $\mathscr{L}$.
2. For any $a_0, a_1 \in L, b_0, b_1 \in I, (a_0 \vee b_0) \wedge (a_1 \vee b_1) = ((a_0 \vee b_0) \wedge a_1) \vee ((a_0 \vee b_0) \wedge b_1) = (a_0 \wedge a_1) \vee (b_0 \wedge a_1) \vee (a_0 \wedge b_1) \vee (b_0 \wedge b_1)$ for $\mathscr{L}$ is a distributive lattice. Since $\mathscr{I}$ is an ideal of $\mathscr{L}$, $b_0 \wedge a_1, a_0 \wedge b_1, b_0 \wedge b_1 \in I$. Therefore there exists $b = (b_0 \wedge a_1) \vee (a_0 \wedge b_1) \vee (b_0 \wedge b_1) \in I$ such that $(a_0 \vee b_0) \wedge (a_1 \vee b_1) = (a_0 \vee a_1) \vee b$.

Consequently, the first condition holds. Since the second condition holds by the assumption, $\mathscr{I}$ is a normal subalgebra of $\mathscr{L}$ with respect to "$\vee$".

The importance of the concept of normal subalgebras comes from the fact that we have a congruence relation induced by it. The following proposition, corollary, and theorem are analogous to those of normal subgroups.

Proposition 3.1 *Let $\mathscr{A} =< A; \{@\} \cup F >$ be a semigroup on "@" and $\mathscr{B} =< B; \{@\} \cup F >$ a normal subalgebra of $\mathscr{A}$. Let us define a binary relation $\sim_{\mathscr{B}}$ on A by*

$$a \sim_{\mathscr{B}} a' \Leftrightarrow a@B = a'@B \text{ for any } a, a' \in A$$

Then $\sim_{\mathscr{B}}$ is a congruence relation on $\mathscr{A}$, where $a@B = \{a@b \mid b \in B\}$.

From now on, all of the proofs of propositions, theorems, and corollaries are placed at the appendix for the sake of readability.

Similar to the case of groups, we have a quotient algebra of a given algebra by a congruence relation induced by a normal subalgebra.

Corollary 3.1 *Let $\mathscr{A} =< A; \{@\} \cup F >$ be a semigroup on "@" and $\mathscr{B} =< B; \{@\} \cup F >$ a normal subalgebra of $\mathscr{A}$ with respect to "@". Then $\mathscr{A}/\sim_{\mathscr{B}}$ is a quotient algebra of $\mathscr{A}$ by $\sim_{\mathscr{B}}$, where $\sim_{\mathscr{B}}$ is defined in Proposition 3.1.*

The following theorem is one of the main theorems of this essay.

Theorem 3.1 *Let* $\mathscr{A} =< A; \{@\} \cup F >$ *be a semigroup on "@" and* $\mathscr{C} =< C; \{@\} \cup F >$ *a subalgebra of* $\mathscr{A}$*. If there exists a normal subalgebra* $\mathscr{B} =< B; \{@\} \cup F >$ *of* $\mathscr{A}$ *such that*

1. *for any* $c, c' \in C$, $c \sim_{\mathscr{B}} c' \Rightarrow c = c'$,
2. $A \subset \{b@c \mid b \in B \wedge c \in C\}$

then $\mathscr{A}/\sim_{\mathscr{B}} \cong \mathscr{C}$ *holds.*

By the above theorem, we can identify a subalgebra of a given algebra with a quotient algebra, which can be considered as a "component" if we can find a normal subalgebra of a given algebra satisfying the conditions.

The following theorem shows a sufficient condition under which a given semigroup on "@" can be identified as a subdirect product of two semigroups on "@" (Graetzer, 1968).

Theorem 3.2 *Let* $\mathscr{A}_0 =< A_0; \{@\} \cup F >$ *and* $\mathscr{A}_1 =< A_1; \{@\} \cup F >$ *be semigroups on "@". Let* $A \subset A_0 \times A_1$ *be such that* $\mathscr{A} =< A; \{@\} \cup F >$ *is a subdirect product of* $\mathscr{A}_0$ *and* $\mathscr{A}_1$*. Let* Θ_0 *and* Θ_1 *be defined by* $\Theta_i = Ker(p_i)$ *for* $i = 0, 1$,*where* p_0 *and* p_1 *are projections defined on* $A_0 \times A_1$ *restricted on A, where* $Ker(p_i) = \{(a, a') \mid p_i(a) = p_i(a') \text{ for } a, a' \in A\}$.

Then the followings hold:

1. $\mathscr{A}/\Theta_0 \cong \mathscr{A}_0$ *and* $\mathscr{A}/\Theta_1 \cong \mathscr{A}_1$
2. $\Theta_0 \wedge \Theta_1 = \omega$*, where* ω *is the identity relation on A*

Conversely, the following theorem holds (Graetzer, 1968).

Theorem 3.3 *Let* $\mathscr{A} =< A; \{@\} \cup F >$ *be a semigroup on "@". Let* Θ_0 *and* Θ_1 *be congruence relations on* $\mathscr{A}$ *such that* $\Theta_0 \wedge \Theta_1 = \omega$*. Then* $\mathscr{A}$ *is isomorphic to a subdirect product of* $\mathscr{A}/\Theta_0$ *and* $\mathscr{A}/\Theta_1$.

For each $a \in A$*, we define an* $f_a \in A/\Theta_0 \times A/\Theta_1$ *as in the following manner:*

$$f_a(0) = [a]_{\Theta_0},\ f_a(1) = [a]_{\Theta_1}$$

Let

$$A' = \{f_a \mid a \in A\} \subset A/\Theta_0 \times A/\Theta_1$$

$\mathscr{A} =< A'; \{@\} \cup F >$ *is a subdirect product of* $\mathscr{A}/\Theta_0$ *and* $\mathscr{A}/\Theta_1$ *and the mapping*

$$\phi : a \mapsto f_a$$

is an isomorphism of $\mathscr{A}$ *onto* $\mathscr{A}$.

3.4 Application for Input–Output Systems

In this section, let us apply the previous discussion for input–output systems (Mesarovic and Takahara, 1989).

Definition 4.1 (Input–Output System) Let X and Y be sets and S a subset of the product of X and Y, that is, $S \subset X \times Y$. Then S is called an **input–output system**, X is called the input set of S, and Y is called the output set of S. From now on, we assume $X = dom(S) = \{x|(\exists y \in Y)((x, y) \in S)\}$ and $Y = cod(S) = \{y|(\exists x \in X)((x, y) \in S)\}$.

The following concept of equivalence relations is quite important if we consider input–output systems.

Definition 4.2 (Input–Output Compatibility for Equivalence Relations) Let $S \subset X \times Y$ be an input–output system with the input set X and the output set Y. Suppose that $\Theta \subset S \times S$ is an equivalence relation on S. Then Θ is called to be **input–output compatible** if the following condition holds:

Both of $\Theta_X = \{(x, x') \mid (\exists y, y' \in Y)((x, y)\Theta(x', y'))\}$ and $\Theta_Y = \{(y, y') \mid (\exists x, x' \in X)((x, y)\Theta(x', y'))\}$ are equivalence relations on X and Y, respectively. In this case, Θ_X and Θ_Y are called **induced equivalence relations** on X and Y based onΘ.

If an equivalence relation Θ on S is input–output compatible, then we can induce equivalence relations Θ_X and Θ_Y on X and Y, respectively by the Definition 4.2. Conversely, we can construct an input–output compatible equivalence relation on S by equivalence relations defined on X and Y that are identical to the induced equivalence relations by Θ.

Proposition 4.1 *Let $S \subset X \times Y$ be an input–output system with the input set X and the output set Y. Suppose that $\Theta_X \subset X \times X$ and $\Theta_Y \subset Y \times Y$ are equivalence relations on X and Y, respectively. Let us define a binary relation Θ on S by*

$$(x, y)\Theta(x', y') \Leftrightarrow (x, x') \in \Theta_X \wedge (y, y') \in \Theta_Y$$

Then Θ is an input–output compatible equivalence relation on S. Furthermore, the corresponding Θ'_X and Θ'_Y induced by Θ following Definition 4.2 are identical with Θ_X and Θ_Y.

Let us next define a complex system that is a subdirect product of input–output systems.

Definition 4.3 (Complex Input–Output System) Let $\boldsymbol{S} = \{S_i \mid S_i \subset X_i \times Y_i, i = 1, \ldots, n\}$ be a class of input–output systems. Then a **complex input–output system** over the class $\boldsymbol{S}$ is a subdirect product of $\boldsymbol{S}$. Each $S_i \in \boldsymbol{S}$ is called a **component**

subsystem of S. S can be considered as an input–output system over $X' \subset X_1 \times \cdots \times X_n$ and $Y' \subset Y_1 \times \cdots \times Y_n$ by identifying

$$((x_1, y_1), \ldots, (x_n, y_n)) \in S_1 \times \cdots \times S_n$$

with

$$((x_1, \ldots, x_n), (y_1, \ldots, y_n)) \in X \times Y.$$

In this section, we will focus on input–output systems that are semigroups on "@". Let us next define the concept as follows.

Let $\mathcal{X} =< X; \{@\} \cup F >$ and $\mathcal{Y} =< Y; \{@\} \cup F >$ be semigroups on "@" of the same type. Then the product of $\mathcal{X}$ and $\mathcal{Y}$, $\mathcal{X} \times \mathcal{Y} =< X \times Y; \{@\} \cup F >$ is also a semigroup on "@". In this case, the following holds.

Corollary 4.1 *Let $\mathcal{X} =< X; \{@\} \cup F >$ and $\mathcal{Y} =< Y; \{@\} \cup F >$ be semigroups on @ of the same type. Suppose that $S \subset X \times Y$ is closed under $\{@\} \cup F$. Then $\mathcal{S} =< S; \{@\} \cup F >$ is a subalgebra on $\mathcal{X} \times \mathcal{Y}$.*

From now on, we focus on an input–output system $\mathcal{S} =< S; \{@\} \cup F >$ that is a subdirect product of $\mathcal{X}$ and $\mathcal{Y}$.

Proposition 4.2 *Both of $dom(S)$ and $cod(S)$ are semigroups on "@".*

Definition 4.4 (Input–Output Compatible Normal Subalgebra) Let $\mathcal{S} =< S; \{@\} \cup F >$ be an input–output system on $\mathcal{X} \times \mathcal{Y}$ and $\mathcal{T} =< T; \{@\} \cup F >$ a normal subalgebra of $\mathcal{S}$. Let us define a binary relation $\sim_{\mathcal{T}}$ on S by

$$(x, y) \sim_{\mathcal{T}} (x', y') \Leftrightarrow (x, y)@T = (x', y')@T \text{ for any } (x, y), (x', y') \in S$$

Then $\sim_{\mathcal{T}}$ is a congruence relation on $\mathcal{S}$, where $(x, y)@T = \{(x, y)@(x', y') \mid (x', y') \in T\}$ as shown in Proposition 3.1. In this case, we call $\mathcal{T}$ as **input–output compatible normal subalgebra** of $\mathcal{S}$ if $\sim_{\mathcal{T}}$ is input–output compatible.

If $\mathcal{T}$ is an input–output compatible normal subalgebra, the corresponding congruence relation $\sim_{\mathcal{T}}$ induces equivalence relations $\sim_{\mathcal{T}_X}$ on X and $\sim_{\mathcal{T}_Y}$ on Y, respectively, as Definition 4.4. Furthermore, it is easily seen that those are congruence relations on $\mathcal{X}$ and $\mathcal{Y}$, respectively.

Then the following theorems are directly obtained from Theorem 3.1 to Theorem 3.3.

Theorem 4.1 *Let $\mathcal{X} =< X; \{@\} \cup F >$ and $\mathcal{Y} =< Y; \{@\} \cup F >$ be semigroups on "@". And $\mathcal{S} =< S; \{@\} \cup F >$ is an input–output system of $\mathcal{X}$ and $\mathcal{Y}$. Suppose that $\mathcal{U} =< U; \{@\} \cup F >$ is a subalgebra of $\mathcal{S}$ and $\mathcal{T} =< T; \{@\} \cup F >$ of $\mathcal{S}$ is an input–output compatible normal subalgebra satisfying the following conditions:*

1. *For any $(x, y), (x', y') \in U$, $(x, y) \sim_{\mathcal{T}} (x', y') \Rightarrow (x, y) = (x', y')$*
2. *$S \subset \{(x, y) \cdot (x', y') \mid (x, y) \in T \wedge (x', y') \in U\}$,*

then $\mathcal{S} / \sim_{\mathcal{T}} \cong \mathcal{U}$ holds.

Since a subalgebra $\mathscr{U}$ of $\mathscr{S}$ can be considered as a subsystem and a quotient algebra of $\mathscr{S}$ divided by $\sim_{\mathscr{T}}$ can be considered as a component subsystem of $\mathscr{S}$, respectively, the above theorem shows the condition under which we can identify a subsystem and a component subsystem of a given system.

The following theorem is the corresponding theorem for input–output systems to Theorem 3.2.

Theorem 4.2 *Let* $\mathscr{X}_i =< X_i; \{@\} \cup F >$ *and* $\mathscr{Y}_i =< Y_i; \{@\} \cup F >$ *be semigroups on "@" of the same type for* $i = 0, 1$. *Suppose that* $\mathscr{S}_i \subset \mathscr{X}_i \times \mathscr{Y}_i$ *is an input–output system for* $i = 0, 1$. *Let* $\mathscr{S} =< S; \{@\} \cup F >$ *be a subdirect product of* $\mathscr{S}_0$ *and* $\mathscr{S}_1$. *Let* Θ_0 *and* Θ_1 *be defined by* $\Theta_i = Ker(p_i)$ *for* $i = 0, 1$, *where* p_0 *and* p_1 *are projections defined on* $S_0 \times S_1$ *restricted to S.*

Then the followings hold:

1. $\mathscr{S}/\Theta_0 \cong \mathscr{S}_0$ *and* $\mathscr{S}/\Theta_1 \cong \mathscr{S}_1$
2. $\Theta_0 \wedge \Theta_1 = \omega$

Conversely, the following theorem is the corresponding theorem for input–output systems to Theorem 3.3.

Theorem 4.3 *Let* $\mathscr{S} =< S; \{@\} \cup F >$ *be a subdirect product of* $\mathscr{X} =< X; \{@\} \cup F >$ *and* $\mathscr{Y} =< Y; \{@\} \cup F >$, *where* $\mathscr{X}$ *and* $\mathscr{Y}$ *are semigroups on "@" of the same type. Let* Θ_0 *and* Θ_1 *be congruence relations on* $\mathscr{S}$ *such that* $\Theta_0 \wedge \Theta_1 = \omega$. *Then* $\mathscr{S}$ *is isomorphic to a subdirect product of* $\mathscr{S}/\Theta_0$ *and* $\mathscr{S}/\Theta_1$.

For each $(x, y) \in S$, *we define an* $f_{(x,y)} \in S/\Theta_0 \times S/\Theta_1$ *in the following manner:*

$$f_{(x,y)} = ([(x, y)]_{\Theta_0}, [(x, y)]_{\Theta_1})$$

Let

$$S' = \{f_{(x,y)} \mid (x, y) \in S\} \subset S/\Theta_0 \times S/\Theta_1$$

Then S' *is closed under* $\{@\} \cup F$ *and hence* $\mathscr{S} =< S'; \{@\} \cup F >$ *is a subdirect product of* $\mathscr{S}/\Theta_0 \times \mathscr{S}/\Theta_1$ *and the mapping* $\phi : S \to S'$ *defined by*

$$\phi : (x, y) \mapsto f_{(x,y)}$$

is an isomorphism between $\mathscr{S}$ *and* $\mathscr{S}'$.

3.5 Conclusions

After reviewing fundamental concepts of algebras, we firstly investigate an algebraic structure in which we can identify a subalgebra with its quotient algebra in this essay. The answer is a normal subalgebra of a semigroup on "@". That is, if we can find a subalgebra $\mathscr{B}$ of a semigroup on "@" $\mathscr{A} =< A; \{@\} \cup F >$for a given subalgebra $\mathscr{C}$ of $\mathscr{A}$, which satisfies the conditions, then we have a congruence relation $\sim_{\mathscr{B}}$ on $\mathscr{A}$ induced by $\mathscr{B}$ and $\mathscr{C}$ can be identified with a quotient algebra $\mathscr{A}/\sim_{\mathscr{B}}$.

Secondly, we extended our discussion to input–output systems that are semigroups on "@". In this case, we find the condition under which we can identify a subsystem of a given input–output system as its component subsystem.

One of our future research should be to find a concrete and meaningful example to which we can apply the discussion in this essay.

Appendix: Proofs

Proof of Proposition 3.1 It is obvious that $\sim_{\mathscr{B}}$ is an equivalence relation on A for it is defined by equality.

Let $f_\gamma \in F$ and $a_0, \ldots, a_{n_\gamma-1}, a'_0, \ldots, a'_{n_\gamma-1} \in A$ be arbitrary. Suppose that $a_i \sim_{\mathscr{B}} a'_i$ holds for $i = 0, \ldots, n_\gamma - 1$. Then there exist $b_0, \ldots, b_{n_\gamma-1}$ such that $a_i = a'_i@b_i$ for $i = 0, \ldots, n_{\gamma-1}$ by the second condition of the definition of normal subalgebras. Therefore for any $b \in B$,

$$\begin{aligned}
&f_\gamma(a_0, \ldots, a_{n_\gamma-1})@b \\
&= f_\gamma(a'_0@b_0, \ldots, a'_{n_\gamma-1}@b_{n_\gamma-1})@b \\
&= (f_\gamma(a'_0, \ldots, a'_{n_\gamma-1})@b')@b \\
&= f_\gamma(a'_0, \ldots, a_{n_\gamma-1})@(b'@b)
\end{aligned}$$

hold for some $b' \in B$ by the first condition of the definition of normal subalgebras and the associativity of "@". Hence

$$f_\gamma(a_0, \ldots, a_{n_\gamma-1})@b \in f_\gamma(a'_0, \ldots, a'_{n_\gamma-1})@B$$

holds. The converse also holds and hence,

$$f_\gamma(a_0, \ldots, a_{n_\gamma-1}) \sim_{\mathscr{B}} f_\gamma(a'_0, \ldots, a'_{n_\gamma-1})$$

Consequently, the substitution property holds and hence $\sim_{\mathscr{B}}$ is a congruence relation on $\mathscr{A}$. □

Proof of Corollary 3.1 Since $\sim_{\mathscr{B}}$ is a congruence relation by Proposition 3.1, it is obvious that $\mathscr{A}/\sim_{\mathscr{B}}$ is a quotient algebra in Sect. 3.3. Let $\mathscr{B} =< B; \{@\} \cup F >$ be a normal subalgebra of $\mathscr{A}$ satisfying the condition. Then $\mathscr{A}/\sim_{\mathscr{B}}$ is a quotient algebra of $\mathscr{A}$ by Corollary 3.1.

Let $i : \mathscr{C} \rightarrow \mathscr{A}$ and $e : \mathscr{A} \rightarrow \mathscr{A}/\sim_{\mathscr{B}}$ be the insertion and the natural projection, respectively. Then it is easily seen that $h : \mathscr{C} \rightarrow \mathscr{A}/\sim_{\mathscr{B}}$ defined by $h = e \cdot i$ is a homomorphism.

Let $[a] \in A/\sim_{\mathscr{B}}$ be arbitrary. Since $a \in A$ and $A \subset C@B$, there exists $c \in C$ such that $[a] = c@B$. Therefore $h(c) = c@B = [a]$ holds and hence h is surjective.

Let $c, c' \in C$ be arbitrary. Suppose that $h(c) = h(c')$. Then $c \sim_{\mathscr{B}} c'$ holds. Therefore $c = c'$ by the assumption. Hence h is injective.

Consequently, h is an isomorphism and hence $\mathscr{C} \cong \mathscr{A}/\sim_{\mathscr{B}}$. □

Proof of Theorem 3.1 Let $\mathscr{B} =< B; \{@\} \cup F >$ be a normal subalgebra satisfying the conditions. Then $\mathscr{A}/\sim_{\mathscr{B}}$ is a quotient algebra by Corollary 3.1.

Let $i : \mathscr{C} \to \mathscr{A}$ and $e : \mathscr{A}/\sim_{\mathscr{B}}$ be the insertion and the natural projection, respectively. Then it is easily seen that $h : C \to A/\sim_{\mathscr{B}}$ defined by $h = e \cdot i$ is a homomorphism.

Let $[a] \in A/\sim_{\mathscr{B}}$ be arbitrary. Since $a \in A$ and $A \subset C@B$, there exists $c \in C$ such that $[a] = c@B$. Therefore $h(c) = c@B = [a]$ holds and hence h is surjective.

Let $c, c' \in C$ be arbitrary. Suppose that $h(c) = h(c')$. Then $c \sim c'$ holds and hence $c = c'$ by the assumption. Consequently, h is an isomorphism and hence $\mathscr{A}/\sim_{\mathscr{B}} \cong \mathscr{C}$.

□

Proof of Proposition 4.1 Suppose that $\Theta\_X \subset X \times X$ and $\Theta_Y \subset Y \times Y$ are equivalence relations on X and Y, respectively. Let us define a binary relation Θ on S by

$$(x, y)\Theta(x', y')) \Leftrightarrow x, \Theta_X x' \wedge y\Theta_Y y'$$

Then it is easily shown that Θ is an input–output compatible equivalence relation on S. Furthermore, Θ_X and Θ_Y are identical with those equivalence relations induced by Θ in the Definition 4.2. □

Proof of Proposition 4.2 Let $x, x' \in dom(S)$ be arbitrary. Then there exist $y, y' \in Y$ such that $(x, y), (x', y') \in S$. Since $\mathscr{S} =< S; \{@\} \cup F >$ is a subalgebra of $\mathscr{X} \times \mathscr{Y}$, $(x, y)@(x', y') \in S$ holds. Consequently, $x@x' \in dom(S)$ holds. It is similarly proved for the case of $cod(S)$. □

Proof of Corollary 4.1 It is obvious by the definition. □

References

Checkland, P. B. (1981). *Systems thinking, systems practice*. John Wiley & Sons.

Graetzer, G. (1968). *Universal algebra*. D. Van Nostrand Company, Ltd.

Iijima, J. (1996). Equivalence of subsystems and components. *International Journal of General Systems*, *24*(3), 271–279.

MacLane, S., & Birkoff, G. (1974). *Algebra*. Chelsea Pub. Co.

Mesarovic, M. D., & Takahara, Y. (1989). *Abstract systems theory*. Springer-Verlag.

Wymore, W. (1976). *Systems engineering methodology for interdisciplinary teams*. John Wiley & Sons.

Chapter 4
Meta-Analysis of Inter-theoretical Relations: Reduction, Realization, and Micro-Macro Relations of Systems

Hiroshi Deguchi

Abstract This chapter focuses on the inter-theoretical relations that are fundamental issues of philosophy of sciences and general systems theory. We introduce the notion of indefinite designators in set-theoretical language for clarifying proper interpretation of concepts between theories. By using indefinite designators, we can define the meta concept of reduction and realization between macro theory and micro theory. An inter-theoretical network constructed by an inter-theoretical proper interpretation by indefinite designators gives an intensional meaning in the network, meaning a semantic holism. The interpretation of indefinite designators in a structure of a set-theoretical universe such as ZFC gives an extensional meaning in a model. If the model is constructed from real-world data by a specific observation method, then the interpretation gives extensional meaning in the real world. Using the framework, we can provide knowledge pragmatics for working systems scientists and philosophers.

Keywords Indefinite designators · Boundary unfolding · Reduction · Realization · Micro-macro relations

4.1 Introduction

4.1.1 Why and How Inter-theoretical Relations

This chapter focuses on "Inter-theoretical Relations," an essential topic in systems sciences. In the first movement of General Systems Science, they focused on establishing relationships between different disciplines by finding a common structure between theories.

H. Deguchi (✉)
Faculty of Commerce and Economics, Chiba University of Commerce, Tokyo, Japan
e-mail: deguchi@cuc.ac.jp

K. Kijima et al. (eds.), *Systems Research I*, Translational Systems Sciences 26,
https://doi.org/10.1007/978-981-16-9937-5_4

This attempt has been considered a failure, but many fruitful research programs have been created in its exploration. Mathematical general systems theory by Y. Takahara is one of the most critical approaches (Mesarovic & Takahara 1975, 1989). His school used a set-theoretical approach to describe systems formally. Mathematical "Category Theory" and "Universal Algebra" are used to describe the relationship between systems, where "Functor" and "Structural Homomorphism" focus on preserving properties from one theory to another. However, this approach has two fundamental issues to describe inter-theoretical relations under the set-theoretical formulation of systems.

First, there are issues related to the ability to express theories. When various theories are expressed mathematically, there exists a limit to their expressive power in the first-order language of Universal Algebra or the formulation of Category Theory. Any theory needs to be characterized by an appropriate language and properties. When we treat a system as a formalized theory, a category, it is difficult to express it adequately in Category Theory or Universal Algebra (first-order structure). The most common way is to express it in set-theoretical language and structure.

It is then necessary to construct a framework of inter-theoretical relations for the theories defined in the set-theoretical language.

The second is to clarify how the various inter-theoretical relations interconnect theories, i.e., categorical knowledge, and what kind of knowledge pragmatics becomes possible. Emergent properties, reduction, micro-macro relations, and other concepts linked to inter-theoretical relations have not been sufficiently investigated mathematically. As a result, vague and sterile conceptual controversies about essential concepts such as reduction, emergent properties, and micro-macro relations continue. Furthermore, essential concepts such as structure, function, and construction used in various social science fields should be discussed with basic concepts such as reduction, emergent properties, and micro-macro relations. However, this approach remains inadequate (Deguchi, 2021).

It is also common in multiple theories to use the same concept. It is necessary to discuss the "meaning" of these concepts in the context of multiple theories and the inter-theoretical relations that connect them. The treatment of meaning is related to an old argument called the Holism of meaning.

For example, the concepts of "compact" and "sequentially compact" in Topology space are not equivalent under the axioms of Topological space. However, both are equivalent in the Metric space. The equivalence means that the relationship between the two properties related to "Topology" differs depending on the theory. The theoretical meanings of the concepts used in the theory of Topological space and the theory of Metric space are different.

In the multi-world semantics of modal logic, it is argued that the valid concept differs depending on the target world. If the theory written in the set-theoretical language can be treated in the many-worlds semantics of modal logic, it corresponds to the modal logic of S4. However, it is impossible to define the worlds and their relationship as the object of modal logic from the set-theoretical definition of theories. There is no syntactical framework for associating theories on a set-theoretical language. On the other hand, the relations between theories are limited to describing by using set-theoretical structures as the worlds and defining the relations between multiple worlds using homomorphisms in the set-theoretical universe. Structural homomorphisms can only preserve limited properties between theories. The inter-theoretical relationship that "metric space is topological space" cannot be described by structural homomorphism.

Behind these problems, there exists an issue of how constants are interpreted and treated in the language.

Assuming that (Ω, Θ) is a topological structure on ZFC, which is a model of set-theoretical language, then Ω and Θ cannot be treated as constant terms when the axiom of this structure is given mathematically because a constant in a set-theoretical language needs to be interpreted as the only fixed object in the set-theoretical universe.

Within the scope of first-order predicate logic, languages can interpret different sets Ω and relationships and functions on Ω, where the constant can be interpreted as the only fixed object on Ω. However, in set-theoretical languages, one universe must be defined in principle. For example, assuming that ZFC is a set-theoretical universe, various mathematical structures can be defined as sets belonging to the universe.

The Set of Real Numbers R, interpreted as a unique object in the ZFC universe, can be treated as a constant in theory. However, when dealing with the theory of topological space, we cannot define unique elements corresponding to Ω and Θ in the ZFC universe. Therefore, later, we will introduce the concept of "Indefinite Designators," which is neither a constant nor a variable and whose referent differs depending on the interpretation, as an extension of the language of theory. By considering Ω and Θ as indefinite designators, we can interpret Ω and Θ in theory for various topological structures and confirm that the interpretation satisfies the axioms of the topological space.

4.1.2 Philosophical Issues of Inter-theoretical Relation

Philosophically, inter-theoretical relations are based on "predicative judgements" between categorical knowledge (Allwood et al., 1977; Husserl, 2012). The predicative judgement "*A* IS *B*" means that the categorical object "*A*" is in specific relation to the categorical object "*B*." There has been much debate about what this relation IS means.

In natural language, when we say "Aristotelēs IS human" or "Dog IS mammal," this indicates the relationship between the proper noun and the common noun, and between the common noun and the common noun, respectively. The IS indicates some kind of inclusion relation, but there is no unique standpoint on how it should be logically expressed.

If we think of a generic noun as giving an extensional definition for the range that the generic noun denotes in the target domain (Ω), then "Dog IS mammal" denotes the relation dog $\subseteq$ mammal, where the set of dogs is included in the set of mammals.

However, since proper nouns imply a definite term (Constant) in the object domain, "Aristotelēs IS mammal" expresses the relationship that Aristotelēs belongs to the set of mammals, i.e., Aristotelēs $\in$ Mammal.

On the other hand, if nouns are predicates that denote categories, then "Dog IS mammal" denotes the proposition $\forall x$ Dog(x) $\rightarrow$ Mammal(x) in the target domain (Ω).

In contrast, Aristotelēs(x) must satisfy $\exists! x$ Aristotelēs(x), since proper names are definite terms of the domain. If we introduce the definite term $a = \varepsilon x$ Aristotelēs(x) to represent Aristotle, then "Aristotelēs IS human" is $\forall x$ Aristotelēs (x) $\rightarrow$ Mammal (x), and also represents the proposition Mammal(a).

This standard philosophical account is based on first-order predicate logic (Hempel, 1970). However, the theories we deal with cannot be adequately described by first-order predicate logic and require a set-theoretical language. Therefore, explanations by predicative judgement based on first-order predicate logic are useless.

The descriptive power of a language for representing theories is a significant issue when discussing how to represent theories that represent some categorical knowledge, how to represent relations between those theories that correspond to predicative judgments, and what is revealed by the relations between theories.

The relationship between models and axiomatic systems is simple in the scope of first-order predicate logic. The model is represented as a structure on the whole set Ω, as (Ω, $R11$, $R12$, ..., $R21$, $R22$, ...) where $R1i$ and $R2j$ denote the definite terms on Ω and the binary relations on Ω, respectively. Similarly, N-term relations can be introduced. For example, (Ω, $\leq$) denotes an ordinal structure, and "$\leq$" is a binomial relation, denoting "$(x, y) \in \leq$" or "$x \leq y$".

Here, the theory is an axiomatic system described in the language of first-order predicate logic. In order to describe a theory, the language of first-order predicate logic must define the theory's specific fixed terms, binomial relations, and so on. The axioms of a theory are a set of statements, or closed formulas, of first-order predicate logic. An axiom is a statement that is presupposed to hold in the target theory.

Example 4.1 Partial Order Theory Variable = $\{x, y, z, ...\}$

First-order binary relation symbols = $\{\leq\}$

The set Σ of first-order closed sentences using variables and binary relation symbols is called the axiomatic system of the theory.

The axioms of partial order theory are as follows.

$\Sigma = \{\Psi 1, \Psi 2, \Psi 1\}$
$\Psi 1 \equiv \forall x \in \Omega\ x \leq x$
$\Psi 2 \equiv \forall x,y \in \Omega\ x \leq y$ and $y \leq x$ then $x = y$
$\Psi 3 \equiv \forall x,y,z \in \Omega \quad \Omega\ x \leq y$ and $y \leq z$ then $x \leq z$

If $\leq$ is reflexive and transitive, then $\leq$ is called a preorder on Ω.

If $\leq$ is preorder and antisymmetric, then $\leq$ is called partial order on Ω.

Σ|- φ denotes that an axiomatic system Σ proves the statement.

The relationship between an axiom system Σ and a first-order mathematical structure is called an interpretation. If a sentence is valid in a structure, then the interpretation is called satisfied, denoted by $(\Omega, \leq)|{=}\varphi$.

In first-order predicate logic, the sentences of a theory are interpreted on the structure. The structure, where Ω is the model's domain, and Q is a binary relation on Ω, is denoted by (Ω, Q).

Let $\Omega = \{a, b, c, d\}$, and $Q = \{(a, a), (b, b), (c, c), (d, d), (a, b), (a, c), (a, d), (b, c), (b, d)\}$.

If we interpret the binary predicate "$\leq$" as a binary relation Q on the domain Ω of the model, then all the axioms of partial order theory hold on this structure.

Suppose that the constants, predicates, and sentence Ψ of a first-order language are interpreted in the structure (Ω, $R11$, $R12$, ..., $R21$, $R22$, ...). When a sentence Ψ is satisfied on a structure (Ω, $R11$, $R12$, ..., $R21$, $R22$, ...), we call the sentence Ψ satisfied.

It is expressed as follows.

$$(\Omega, R11, R12, \ldots, R21, R22, \ldots) \models \Psi$$

The axiom Σ of partial order theory is satisfied in (Ω, Q) by interpreting the binary predicate "$\leq$" as a binary relation Q on the domain Ω of the model. It can be expressed as $(\Omega, Q)|{=}\Sigma$.

In this way, mathematical structures are used to give an extensional meaning to first-order languages. There, the dualism of syntax and semantics is used, where proofs are given on syntax (representation of theories using axioms), and meaning is given by semantics (interpretation of sentences modeled on mathematical structures). In this dualism, we can discuss the satisfaction of a theory in a model, but it is hard to discuss inter-theory relations.

In this dualism, we can discuss the satisfaction of a theory in a model, but it is challenging to discuss inter-theoretical relations because the inter-theoretical relation is a multi-level relation among syntactical theories. However, the interpretation between theory and structure is a dualistic relation between syntax and semantics.

In mathematical theory, the objects of semantic interpretation are called mathematical structures, such as groups, order structures, and topological spaces. Axioms over first-order predicate logic characterize group theory. If a theory is expressed in first-order predicate logic, then the relation between theory and structure is discussed as a satisfaction relation, while the relation between mathematical structures is discussed in terms of isomorphism or homomorphism. Universal Algebra is a general theory of mathematical structures based on first-order languages (Graetzer, 1979).

However, the range of theories that can be expressed in first-order predicate logic is limited. Many theories in the natural sciences cannot be described in a first-order language. Hence, the primary language in modern mathematics is a set-theoretical one, and axioms characterizing the set-theoretical language (e.g., ZFC) are assumed. Then, a set-theoretical structure is defined for sets belonging to the set-theoretical universe (e.g., the ZFC universe). The set-theoretical structure is a mathematical object used in modern mathematics. The structure of first-order predicate logic can also be included in this set-theoretical structure.

Interpretation and satisfaction between theory and structure using set-theoretical language have different characteristics than first-order predicate logic. For example, let us consider the structure of a topological space. Let (Ω, Θ) be a structure of topological space, and assume that $\Theta \subseteq \mathrm{Power}(\Omega)$ in the topological space. Then, $\varphi \in \Theta, \forall x, y \in \Theta\ x \bigcap y \in \Theta, \forall z \subseteq \Theta \cup z \in \Theta$ holds.

To describe the axioms of the topological space in the set-theoretical language, let Ω, φ, and Θ be constants.

$$\Sigma = \{\Psi 1, \Psi 2, \Psi 1\}$$
$$\Psi 1 \equiv \varphi \in \Theta, \Omega \in \Theta$$
$$\Psi 2 \equiv \forall x, y \in \Theta \rightarrow x \bigcap y \in \Theta$$
$$\Psi 3 \equiv \forall z \subseteq \Theta \cup z \in \Theta$$

Moreover, the satisfaction relation is shown for the set-theoretical universe V by giving interpretations of the constants Ω, φ, and Θ in V.

We deal with various topological spaces as mathematical structures. Since the empty set φ is unique in ZFC, it is common to all topological spaces. However, Ω and Θ are given differently in several structures, satisfying the axioms of topological spaces. This means that unlike in the model of first-order languages, where constants can be interpreted differently for each Ω, the set-theoretic universe is unique, and constants are required to be interpreted uniquely in the set-theoretical universe.

To avoid this aporia, the concept of "Indefinite Designers" will be introduced in the next section.

4.2 Inter-theoretical Relations Based on Indefinite Designators in Set Theory

4.2.1 Indefinite Designators

The conventional approach to inter-theoretical relations admitted the dualism of syntax and semantics; formal theories are written by symbols without meanings and need models to acquire meanings. However, Bourbaki's work showed that even mathematical theories could not be described under such a clear-cut dualism (Bourbaki, 2008). For instance, whereas groups are models of the first-order theory of group, topological spaces cannot be considered models of any purely syntactical theory of topology.

A proposal to circumvent this difficulty was to consider every theory as a semantic entity. Mathematicians describe the theory of topology by directly defining what topological spaces are. However, it is difficult to discuss inter-theoretical relations from this approach since this excludes model theoretical devices based on dualism. Thus we provide a new framework for discussing inter-theoretical relations that enables us to restore the dualism. For this purpose, we introduce a new linguistic category called indefinite designators into the language of set theory.

This concept is necessary for distinguishing notions proper to the theory under consideration from mathematical notions common to every theory, such as numbers and functions. The proper notions are expressed syntactically by indefinite designators and can be interpreted as arbitrary sets in a fixed universe of set theory, while the common mathematical notions have the standard interpretations in the same universe of set theory.

For a framework to discuss interrelations among various parts of our knowledge, this extension of the set-theoretical language can be an inevitable consequence from the conventional view in which set theory is a unified conceptual basis for scientific theories.

4.2.2 Language of Set Theory

Let V be the universe of sets. We assume that V satisfies the ZFC set theory. Let $L[\in]$ be the language for first-order theory with equality having a binary predicate symbol "∈" and no constant symbols. This is usually called the language of set theory (Takeuti & Zaring, 1982). For every $u \in V$, we introduce the corresponding constant symbol $c[u]$. We denote by $L[\in](V)$ the language $L[\in]$ augmented by the class of symbols $\{c[u] | \ u \in V\}$.

For every closed formula Φ in $L[\in](V)$, the satisfaction relation $<V,\in>|=\Phi$ is defined by the following recursive rules:

1. $<V,\in>|=c[u] = c[v]$ if and only if u and v are identical.
2. $<V,\in>|=c[u] \in c[v]$ if and only if u is an element of v.
3. $<V,\in>|=\neg\Phi$ if and only if $<V,\in>|=\Phi$ does not hold.
4. $<V,\in>|=\Phi \vee \Psi$ if and only if $<V,\in>|=\Phi$ holds or
5. $<V,\in>|=\Psi$ holds.
6. $<V,\in>|=(\exists x)\Phi(x)$ if and only if there exists some u in V such that $<V,\in>|=\Phi(c[u])$ holds.

For any closed formula Φ including logical symbols such as $\wedge$, $\Rightarrow$, and $\forall$, we will use a closed formula Ψ logically equivalent to Φ without those symbols to define the relation $<V,\in>|=\Phi$ to be equivalent with the relation $<V,\in>|=\Psi$.

A closed formula Φ in $L[\in]$ is said to be true in the standard interpretation, if $<V,\in>|=\Phi$ holds.

Since we assume that V satisfies the ZFC set theory if $\Phi(x1,...,xn)$ is a formula in $L[\in]$ and provable from axioms of the ZFC set theory, for any $u1,..., un \in V$ we have $<V,\in>|=\Phi(c[u1],...,c[un])$.

In what follows, we assume that for well-known symbols for sets, functions, and relations definable in the ZFC set theory, we can augment the language by adding those symbols to the language $L[\in]$ as constant symbols and extend the axioms by adding the definitions of those symbols to the axioms of the ZFC set theory. For instance, we can add such symbols as φ for the empty set, 0,1,2,... for natural numbers, R for the real number field, $\cup$ for the join of sets, $\subseteq$ for the inclusion relation by the above manner. Nevertheless, the above extension is a conservative one, and to any formula, in the augmented language, we have an equivalent formula without having those new symbols. Thus, we will take those symbols as abbreviations to be eliminated by definition and not consider the extended theory in our formal treatment.

4.3 Problems of IS Relations in Inter-theoretical Relations

In mathematicians, one investigates arbitrary $|=$ of axiomatic group theory.

However, it is rather in meta-mathematics that one investigates arbitrary $|=$ of axiomatic set theory. This is related to the fact that set theory can be considered a theory for the unique entity of the universe of sets, but that group theory is regarded as a theory for mathematical structures of groups that can be found throughout. Many mathematical theories and almost every scientific theory are intermediate between them. Those theories assume the uniqueness of the mathematical objects such as individual numbers, individual functions, and set-theoretical operations, and using those mathematical objects, and they aim to express universal structures in mathematics or empirical sciences in the most general

In the conventional model theory, it is taken for granted that predicate calculus is a common basis for all theories and that every mathematical notion is proper for each theory. However, to realize the systematic description of various scientific theories in a unified language, the above considerations suggest the need to include predicate calculus and set theory in the common basis for all theories.

To be more specific, let us consider how to formally express the structural notions such as metric spaces and topological spaces by the language of set theory. It is well known that predicates can express those notions in the language $L[\in]$.

The predicate Met(x) expressing "x is a metric space" is defined in $L[\in]$ as follows.

$$\begin{aligned} \mathrm{Met}(x) \iff (\exists \Omega, \rho)\Big[& x =< \Omega, \rho > \wedge \Omega \neq \phi \wedge \rho : \Omega \times \Omega \to R \\ & \text{and } (\forall y1, y2 \in \Omega)\, \rho\,(y1, y2) \geqq 0 \\ & \text{and } (\forall y1, y2 \in \Omega)\, \rho\,(y1, y2) = 0 \iff y1 = y2 \\ & \text{and } (\forall y1, y2 \in \Omega)\, \rho\,(y1, y2) = \rho\,(y2, y1) \\ & \text{and } (\forall y1, y2, y3 \in \Omega)\, \rho\,(y1, y2) + \rho\,(y2, y3) \geqq \rho\,(y1, y3)\Big] \end{aligned}$$

Similarly, the predicate Top(x) expressing "*x* is a topological space" is defined in $L[\in]$ as follows.

$$\begin{aligned}\mathrm{Top}(x) \iff & (\exists \Omega, \Theta)\Big[x =< \Omega, \Theta > \wedge \Omega \neq \phi \wedge \Theta \subseteq \mathrm{Power}\,(\Omega)\\ & \text{and } \phi \in \Theta \text{ and } \Omega \in \Theta\\ & \text{and } (\forall y1, y2 \in \Theta)\, y1 \cap y2 \in \Theta\\ & \text{and } (\forall y \subset \Theta) \cup y \in \Theta\end{aligned}$$

It is generally accepted that for a given pair of common names "*A*" and "*B*," the truth value of the "IS" relation "An *A* IS *B*" coincides with that of the formula "$(\forall x)[A(x) \Rightarrow B(x)]$" obtained by the predicate $A(x)$ expressing "*x* IS *A*" and the predicate $B(x)$ expressing "*x* IS *B*".

On the other hand, in mathematics, an "IS" relation such as "A metric space is a topological space" is often used effectively. However, this "IS" relation cannot be expressed by the formula "$(\forall x)[\mathrm{Met}(x) \Rightarrow \mathrm{Top}(x)]$" obtained by the predicates Met(*x*) and Top(*x*) formalized as above. In this case, the above formula is false.

A method to avoid this difficulty is to read the predicate Met(*x*) as "*x* is a metric space structure" instead of "*x* is a metric space" and to read the predicate Top(*x*) as "*x* is a topological space structure" instead of "*x* is a topological space." Then, the falseness of the formula "$(\forall x)$ [Met(*x*) $\Rightarrow$ Top(*x*)]" is in accordance with the falseness of the "IS" relation "A metric space structure is a topological space structure" according to our definitions of the notions of those structures.

However, this method cannot escape from the criticism that it takes advantage of the ambiguity of the extensions of the words "metric space structure" and "topological space structure" and fails to express the relation found between the intentions of those words.

For instance, we have two optional definitions of "a topological space structure," one formalized as the set of open subsets and one formalized as the set of neighborhoods, with different extensions. Hence, it is difficult to uniquely determine the extension of the word "topological space structure."

It follows that the formalization of the type "$(\forall x)[A(x) \Rightarrow B(x)]$", which expresses the inclusion relation between the extensions of "*A*" and "*B*," is not suitable for the present situation.

Now, how should we formalize "IS" relations between mathematical structures? Our leading principle to this problem is the assumption that such a structural notion always has a corresponding theory that can be identified with a certain syntactical formal system. Then, the "IS" relation "*A* metric space is a topological space," for example, can be justified by interpreting it as "what the theory of metric spaces can apply to is what the theory of topological spaces can apply to in the corresponding standard method."

4.4 Inter-theoretical Relations Based on Set Theory

4.4.1 Indefinite Designators and Theories over the Set Theory

We introduce a new set of constant symbols, called the **infinite designators**, into the language $L[\in]$. We denote by $L[\in][N]$ and $L[\in][N](V)$ the languages $L[\in]$ and $L[\in](V)$, respectively, augmented by adding the set N of infinite designators.

For a set N of infinite designators and a set A of closed formulas of $L[\in][N]$, the pair (N, A) is called the theory over set theory with set N of infinite designators and axiom system A. In what follows, a theory over set theory will be simply called a theory as long as no confusion may occur. In this case, the language $L[\in][N]$ is called the language of the theory (N,A) and an element of A is called an axiom of the theory (N,A).

A formula Φ of the language of (N, A) is called a theorem of (N, A) if it is provable from axioms of the ZFC set theory and axioms of (N, A).

In this case, we write ZFC, $A|\text{-}\Phi$.

Example 4.2 Theory of topological spaces Σ[Top] Theory of topological spaces $\Sigma[\text{Top}]$ is the theory with a set $\text{NV1} = \{\Omega, \Theta\}$ of infinite designators and the axiom system $A1$ given by the following (1)–(3).

1. $\Theta \subseteq \text{Power}(\Omega) \wedge \Omega \in \Theta \wedge \varphi \in \Theta$.
2. $(\forall X,Y \in \Theta)\, X \cup Y \in \Theta$.
3. $(\forall H \subseteq \Theta) \cup H \in \Theta$.

Example 4.3 Theory of metric spaces Σ[Met] Theory of metric spaces $\Sigma[\text{Met}]$ is the theory with a set $\text{NV2} = \{\Omega',\rho\}$ of infinite designators and the axiom system $A2$ given by the following (1)–(4).

1. $\Omega' \neq \varphi$ and $\rho{:}\Omega' \times \Omega' \rightarrow R$.
2. $(\forall y1,y2 \in \Omega')\rho(y1,y2) \geq 0$.
3. $(\forall y1,y2 \in \Omega')\rho(y1,y2) = 0 \iff y1 = y2$.
4. $(\forall y1,y2 \in \Omega')\rho(y1, y2) = \rho(y2, y1)$.
5. $(\forall y1,y2,y3 \in \Omega')\rho(y1,y2) + \rho(y2,y3) \geq \rho(y1, y3)$.

4.4.2 Interpretation of Theories

Since theory (N, A) is described by the language $L[\in][N]$ for first-order theory with equality having the set of constant symbols N, we have interpretations of the language $L[\in][N]$, as every first-order theory with equality, which is so general to allow nonstandard interpretations of set theory. On the other hand, the language $L[\in]$ has the standard interpretation formulated as before, and mathematical notions in theory (N, A) are commonly interpreted under this standard interpretation.

Thus, among general interpretations of $L[\in][N]$, we take the interpretation of theory (N, A) as those interpretations of $L[\in][N]$ such that the standard interpretation interprets the language $L[\in]$ and that indefinite designators can be interpreted as arbitrary elements of the ZFC universe V.

Formalizing this idea, we define an interpretation of theory (N, A) to be any function I from the set N of indefinite designators to the ZFC universe V.

The satisfaction relation under an interpretation of theory (N, A) is defined by reducing it to the corresponding relation under the standard interpretation for the formula without indefinite designators as follows. For any closed formula with indefinite designators $a1,..., an$, let $I(\Phi)$ be the formula obtained by replacing every indefinite designator ai in Φ by $c[I(ai)]$. Then, formula Φ is said to be satisfied under the interpretation "I" of theory (N, A), if the formula $I(\Phi)$ is true under the standard interpretation, i.e., $<V,\in>|=I(\Phi)$. In this case, we shall write $I|=\Phi$.

4.4.3 *Extensions by Definition of Theories*

Let (N, A) be a theory. For a predicate $\Phi(x)$ with one variable of the language $L[\in][N]$, if $(\exists! y)(\forall x)(x \in y \iff \Phi(x))$ is a theorem of the theory (N, A), then we can introduce new indefinite designators "c" by the following axiom which is called $\alpha[c]$.

$$(\forall x)\,(x \in c \iff \Phi(x)) \quad \alpha\,[c]$$

The extended theory $(N \cup \{c\}, A \cup \{\alpha[c]\})$ is called the extension of the theory (N, A) by $(c, \alpha[c])$.

Example 4.4 Extension by Definition of Metric Space We introduce an indefinite designator $\Theta[\rho]$ for the extension by definition of metric space. NV2ex $= \{\Omega',\rho,\Theta[\rho]\}$. Then we introduce the following axiom to characterize the indefinite designator $\Theta[\rho]$.

$$l\,(\forall X)\,[X \in \Theta\,[\rho] \iff X \subseteq \Omega \wedge (\forall x \in X)\,(\exists \varepsilon \in R)\,(\forall y \in \Omega)\;\rho\,(x, y) < \varepsilon \Rightarrow y \in \mathrm{X}]$$

Σ[Met-ex] denotes the extended theory.

As is well known in the first-order theory, the extension by definition is a conservative extension. Thus the following proposition hold.

Proposition 4.1

1. Let Φ is a formula in $L[\in][N]$. If Φ is a theorem of theory $(N \cup \{c\}, A \cup \{\alpha[c]\})$, then Φ is a theorem of (N, A).
2. For any formula Φ in $L[\in][N \cup \{c\}]$ there exists formula Ψ ZFC $\cup \{c\}|\text{-}\Phi \iff \Psi$ holds.

4.4.4 Interpretation Between Theories

Let (NV1, $A1$) and (NV2, $A2$) be theories. For given theories (NV1, $A1$) and (NV2, $A2$), any one-to-one mapping J from NV1 to NV2 is called an interpretation from (NV1, $A1$) to (NV2, $A2$). Then the interpretation from $L[\in]$[NV1] to $L[\in]$[NV2] is defined as follows.

Let $\Phi(a1,..., an)$ be a formula Φ in $L[\in]$[NV1] and $a1,..., an$ are infinite indefinite designators appearing in Φ. Then the formula $\Phi(J(a1),..., J(an))$ where indefinite designators $a1,..., an$ are substituted by $J(a1),..., J(an)$ is called J-interpretation of Φ.

Example 4.5 Example of interpretation between theories The interpretation J from Σ[Top] to Σ[Met-ex] is defined as follows.

J: NV1 = $\{\Omega,\Theta\} \rightarrow$ NV2ex = $\{\Omega',\rho,\Theta[\rho]\}$
$J0\,(\Omega) = \Omega'$, $J0\,(\Theta) = \Theta[\rho]$

Then the sentence such as "$\forall X,Y \in \Theta\ X \cap Y \in \Theta$" is interpreted on the Metric Space as follows.

$$J\,(\forall X, Y \in \Theta\ X \cap Y \in \Theta) = \forall X, Y \in \Theta\,[\rho] \quad X \cap Y \in \Theta\,[\rho]$$

The relation is shown in the following figure (Fig. 4.1).

4.4.5 Valid Interpretation Between Theories

Let J be an interpretation from theory (NV1, $A1$) to theory (NV2, $A2$). Then J is called a valid interpretation if and only if $\forall\Phi \in A1\ A2$ |-$J(\Phi)$holds.

From the definition, it is clear that the composition of interpretations becomes an interpretation, and the composition of valid interpretations becomes a valid interpretation.

For example, the interpretation J: Σ[Top] $\rightarrow$ Σ[Met-ex] is valid. Because J-interpretation of any axiom in Σ[Top] becomes a theorem of Σ[Met-ex]. The interpretation J represents the "IS" relation "A metric space is a topological space."

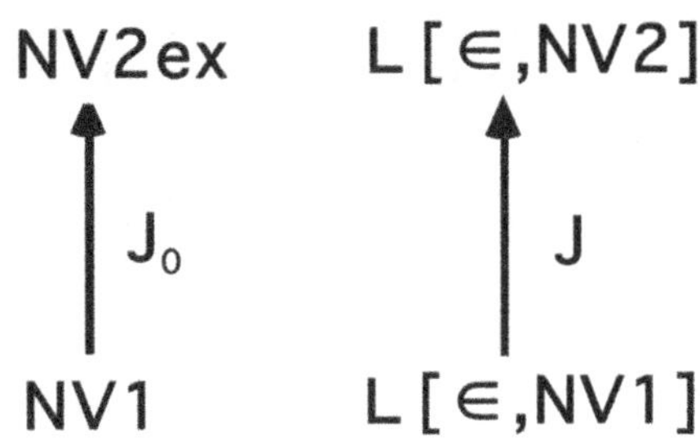

Fig. 4.1 Interpretation between theories

Let mathematical structures A and B are described by $\Sigma[A]$ and $\Sigma[B]$, respectively. If there exists a valid interpretation from $\Sigma[B]$ to a natural extension by definition of $\Sigma[A]$, then the sentence "A is B" is considered to be justified by the interpretation.

Proposition 4.2 Let J be a valid interpretation from a theory ($A1$, NV1) to ($A2$, NV2).

1. For any formula Ψ in $L[\in, \text{NV1}]$,
   if ($A1$, NV1)|-Ψ hold then ($A2$, NV2)|-$J(\Psi)$ hold.
2. For any closed sentence Ψ in $L[\in, \text{NV1}]$,
   if $J(\Psi)$ is proved in ($A2$, NV2), then $\neg\Psi$ is not proved in ($A1$, NV1).

That means Ψ is proved or independent in theory ($A1$, NV1).

4.4.6 Adjoint Extension of Theories

Theorem 4.1 Adjoint extension Let J be a valid interpretation from Γ to Ω and Γex be an extension of Γ by definition.

Then there exist Ω', which is the extension by definition of Ω, and J', which is the extension of J. The Ω' and J' satisfy the following properties, and the relation is shown in Fig. 4.2.

J is the restriction on J' to $L[\Gamma]$.
J' is a valid interpretation from Γex to Ω'.
Then J' and Ω' are called the adjoint extension of Γex (Fig. 4.2).

Example 4.6 Example of adjoint extension Let I be a valid interpretation from Γ to Σex and J be a valid interpretation from Σ to Ωex. Then there exists the valid interpretation J' from Ωex$'$ to Ωex. The relation is shown in the following figure (Figs. 4.3 and 4.4).

[Example 4.7 Construction of Valid Interpretation (Fig. 4.5)]

$$
\begin{array}{ccc}
\Omega & \rightarrow & \Omega' \\
J \uparrow & & \uparrow J' \\
\Gamma & \rightarrow & \Gamma\text{ex}
\end{array}
$$

Fig. 4.2 Adjoint extension

Banach Ring → Topological Space defined on Banach Ring

J ↑ ↑ J'

Banach Space → Topological Space defined on Banach Space

Fig. 4.3 Valid interpretation and its commutative diagram

$\Omega \rightarrow \Omega ex \rightarrow \Omega ex'$

J ↑ ↑ J'

$\Sigma \rightarrow \Sigma ex$

↑ I

Γ

Fig. 4.4 Valid interpretation and its commutative diagram

Banach Space → Metric Space Defined on Banach → Topology Space Defined on Banach

J ↑ ↑ J'

Metric Space → Topology Space Defined on Metric Space

↑ I

Topology Space

Fig. 4.5 Construction of valid interpretation

4.4.7 Abstract Inference, Concrete Inference, Inner Theoretical Inference

Let Σ1, Σ2, and Σ3 be theories. Let I: Σ1 → Σ2 and J: Σ2 → Σ3 be valid interpretations. There are the following three types of methods to show that the sentence Φ in Σ2 is true.

1. **Inference, inner Theoretical Inference**
   To show Σ2|-Φ in Σ2 directly.
2. **Abstract inference**
   If Φ can be pulled back to the abstract theory Σ1 by interpretation I, then it is called an abstract inference by valid interpretation I to show Σ1|-I^{-1}(Φ).
3. **Concrete inference**
   Let J be a valid interpretation from Σ2 toΣ3. Then it is called a concrete inference to show Σ3|-J(Φ).

The inner theoretical inference is a usual way of inference. The abstract inference is possible to execute if Φ can be pulled back. If Σ1|-I^{-1} (Φ) hold, then Σ2|-Φ hold where $I{\cdot}I^{-1}$ (Φ) = Φ. In concrete inference, the proof is limited because even if Σ3|-J(Φ) is shown then it only guarantees that Φ is provable or independent in Σ2.

4.4.8 Pullback of Concepts and Inter-theoretical Equivalence

Three concepts are used in the metric space such as (1) compact, (2) sequentially compact, (3) complete and totally bounded. These three concepts are defined as one variable predicate and are equivalent with each other on metric spaces. Nevertheless, it is impossible to pull the concept of complete and totally bounded back to topological spaces. It is possible to pull the concepts of compact and sequentially compact back to topological spaces. However, the concepts are not equivalent with each other.

From our hierarchical analysis of theories, we can analyze the change of concepts' connotation, such as the change of equivalence among concepts.

In modern mathematics, we tried to construct an abstract theory on which concrete notions are pulled back and analyzed on the abstract theory, especially in functional analysis.

Our analysis clarifies the meaning of these abstractions in modern mathematics.

4.5 Two Dogmas of Reductionism

4.5.1 Two Dogmas of Empiricism and Semantic Holism

The relationship between the theories described by the set-theoretical language can be appropriately expressed in a syntactical framework by introducing indefinite designators into the axiomatic system described by the set-theoretical language, as in the formulation in the previous section for "Metric Space is Topological Space."

In contrast, the issue of this section is to provide an appropriate formalization for the relationship between micro and macro theories and the meta properties between systems, such as emergent properties, realization, and reduction, which have long been discussed in systems science. This section assumes that formal systems are formulated by theories defined over a set-theoretical language that introduces indefinite designators.

If several theories are treated in a whole network of inter-theoretical relations connected by predicate judgments (IS), then this network becomes a formalization for a "holistic theory of meaning" and gives an intensional meaning to the concept of theory. We call this network a "knowledge network" consisting of inter-theoretical relations.

We have already shown that in the knowledge network of inter-theoretical relations, including Metric Space and Topological Space, the intensional meaning of the concepts "compact" and "sequentially compact" differs depending on the theory. In the network of inter-theory relations, the intensional meaning of the two concepts "compact" and "sequentially compact" differs depending on the theory. This means that the intensional meaning of the concepts changes depending on the theory.

In general, if we introduce an ordering relation between theories, "Σ IS Γ" $\Longleftrightarrow$ "$\Sigma > \Gamma$," with the relations between theories connected by the IS relation "Σ IS Γ," and we call a theory a world, and we regard the world Σ as visible from the world Γ, then the relations between worlds constitute the possible worlds of modal logic S4.

The holistic theory of meaning asserts the meaning of a sentence or word depending on its overall structure. The holistic theory of meaning opposes the determination of meaning by the elementary units.

The holistic theory of meaning is often referred to as the knowledge network theory.

As knowledge expands somehow, it seems natural that the meaning of the sentences and words will gradually change. The knowledge network theory discussed in the previous section can also be regarded as a holistic theory of meaning.

The basic argument about the holistic theory of meaning was given by W.V.O. Quine in "Two Dogmas of Empiricism" as follows (Quine, 1980).

> Modern empiricism has been conditioned in large part by two dogmas. One is a belief in some fundamental cleavage between truths which are *analytic*, or grounded in meanings independently of matters of fact and truths which are *synthetic*, or grounded in fact. The other dogma is *reductionism*: the belief that each meaningful statement is equivalent to

> some logical construct upon terms which refer to immediate experience. Both dogmas, I shall argue, are ill founded. One effect of abandoning them is, as we shall see, a blurring of the supposed boundary between speculative metaphysics and natural science. Another effect is a shift toward pragmatism. [Quine, 1980]

In the logical positivism of philosophy of science, knowledge of natural science is described by First Order Language. Under this assumption, "The reduction from theoretical sentence to the observational sentence" and "Analytic Synthetic Distinction" were discussed.

Our problem is to provide an appropriate formulation for inter-theoretical relations such as reduction and realization between systems, emergent properties, and micro-macro relations. Reductions must be defined not for sentences in first-order predicate logic but as relations between theories described by set theoretical language.

The IS relation through indefinite designators between the theories forms a network between the theories, where the intensional meanings of definite terms, indefinite designators, and sentences are defined. In addition, the extensional meanings of the theories are given by the interpretations of the fixed terms, indefinite designators, and sentences into the mathematical structures of the ZFC Universe.

In this definition, the extensional meaning of a theory is given not by the real world, but by the mathematical structure. In contrast, the interpretation of a theory into the real world is given by the "Experimental Observation Model, which is built on the ground of experiments and observations" (Fig. 4.6).

For example, macro Ohm's law $V(t) = I(t) * R(t)$ has the indefinite designators $V(t)$, $I(t)$, and $R(t)$. Experimental observation $(x(t), y(t), z(t))$ where $x(t)$ denotes observed electric voltage, $y(t)$ denotes observed electrical resistance, and $z(t)$ denotes electric current at time t in the experiments, respectively. Then $(x(t), y(t), z(t))$ denotes an experimental observation model.

Macro Ohm's law "$V(t) = I(t) * R(t)$" in theory can be interpreted on a mathematical model in ZFC Universe or on an experimental observation model, that is also in ZFC Universe and constructed by experimental data from "Real World."

In order to create an experimental observation model from the "Real World," various auxiliary hypotheses and experimental instruments are required.

A method to distinguish experimental observation models from mathematical models, and to investigate the interpretation and satisfaction of theories, has already

Theory → Math. Model as a Math. Structure in ZFC Universe
Model Interpretation ⇅

⇅

Theory → Experimental Observation Model → Real World
Experimental Model Interpretation Experimental Interpretation

Fig. 4.6 Interpretation in math model and experimental observation model

been discussed by Frederick Suppe as the relation between models and observations for first-order predicate logic (Suppe, 1977).

Using a set-theoretical language, we can formulate almost mathematical and physical models as mathematical theories. The issues are the inter-theoretical relations, the interpretation of the theory in its mathematical structure, and the interpretation of the theory in its experimental observation model.

From our point of view, the aporia that if theoretical terms are reduced to observational terms (sense data), then "Observational Theoretical Distinction does not hold" and "Theoretical terms are unnecessary" does not hold. The aporia was a problem in logical positivism,

Theories are given intensional meanings through inter-theoretical relations, and extensional meanings are given through interpretations of mathematical structures in the formal world. Furthermore, if the valid satisfaction of an interpretation to a model constructed from experimental observations can be shown, it implies that the theory can explain the real world.

The assertion is different from the dogma of reductionism that Quine rejects. However, the distinction between "Mathematical Model" and "Experimental Observation Model" provides a kind of "Observational Theoretical Distinction."

Theories are given interpretations on mathematical models, apart from their interpretations on experimentally constructed models. On the mathematical model, we can discuss the conceptual meaning of the theory.

However, to interpret a theory in the real world, it is necessary to construct an "Experimental Observation Model" from "Experimental Observation Data" and show that the theory is valid on this model. If we regard the interpretation of the theory in the real world as "Synthetic Interpretation" and the interpretation of the theory in the mathematical structure as "Analytic Interpretation," then "Analytic Synthetic Distinction" becomes true.

On the other hand, we insist on Semantic Holism, which says that meaning is determined in the overall structure of inter-theoretical relations. In addition, we propose "Knowledge Pragmatism," which relates theories, inter-theoretical relations, mathematical models as mathematical structures, "Experimental Observation Data," and "Experimental Observation Model" with each other to operate various kinds of knowledge from the real world to the mathematical world. This is different from the semantic holism claimed by Quine (1980).

The purpose of logical positivism was to make a demarcation between scientific and non-scientific knowledge by logical reconstruction of scientific knowledge. Nowadays, this research program is considered to have failed (Suppe, 1977). However, no formal framework for the logical analysis of scientific knowledge has been developed to replace it.

We will discuss in concrete terms what a holistic theory of meaning in a knowledge network reveals about scientific knowledge from theories and inter-theoretical relations over set-theoretical languages. This section develops a knowledge network theory based on the intensional and extensional semantics of theories described by set-theoretical languages, including indefinite designators and their inter-theoretical interpretations and interpretations to models.

The knowledge network theory proposes a kind of holistic theory of meaning. However, the framework of this theory is entirely different from Quine's holistic theory of meaning based on his criticism of the distinction between analytic and synthetic sentences and the reduction of theoretical sentences to observational sentences, and from many other discussions of the holistic theory of meaning (Fodor & Lepore, 1992).

4.5.2 Reduction from Macro Theory to Micro Theory

We talked about mathematical knowledge in this section. Set theory and mathematics, depending on axiomatic set theory, are fundamental tools for describing scientific knowledge. Thus our discussion with indefinite designators can be applied to the knowledge structure of natural sciences.

As an example of natural science's knowledge structure, we treat the relation between classical Macro Ohm's law and Micro Ohm's law in electromagnetic theory.

We can formulate the assertion that classical Ohm's law "$V = IR$" can be reduced to Micro Ohm's law "$i = \kappa E$" where "E" is the electric field vector, "i" is the current density vector, and κ is the conductivity. Let Γ be the micro continuum form of Ohm's Law. Let Σ be Macro Ohm's law.

The formula "$i = \kappa E$" is called Micro Ohm's law and is not identical to macro Ohm's law "$V = IR$." Identifying these two laws becomes possible after constructing a specific relation between theories.

We assume that classical Macro Ohm's law is defined on such a special conductor that the length is L, the force of the electric current is parallel to the conductor, the density of electric current "i" is constant (steady state), and the cross-section area of the conductor is S. Under the particular boundary conditions, we introduce the following indefinite designators I, V, R as extensions by definition of Maxwell's law.

$$I = \int Si \cdot n\mathrm{d}S$$

$$V = -\int E \cdot \mathrm{d}L$$

$$R = L/\left(S\kappa\right)$$

From the steady-state condition, the following properties hold.

$$I = S \mid i \mid = S \mid \kappa E \mid, \quad V = \mid E \mid L, \quad L = RS\kappa$$

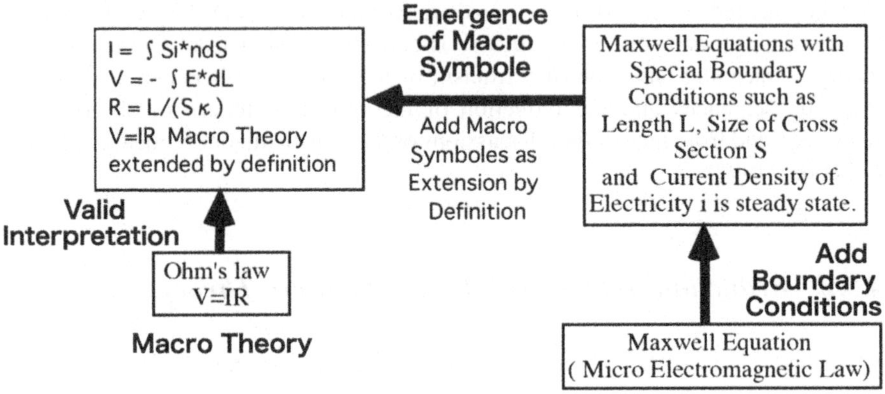

Fig. 4.7 Reduction of classical macro Ohm's law to micro electromagnetic law

Thus $V = |E|L = |E|RS\kappa = RS|\kappa E| = IR$ holds.

Then classical Macro Ohm's law can be reduced to Micro Ohm's law (Maxwell's law). Micro Ohm's law Γ is specified to the theory Γ' by adding the axioms that show the special boundary conditions. We can extend the theory Γ' to Γ'ex by definition such as "$V = -\int E{\cdot}dL$." We introduced indefinite designators I, V, R, and its characterization axioms for these extensions by definition. Then natural interpretation I: $\Sigma \rightarrow \Gamma'$ex exists. "$V = IR$" is proved on Γ'ex. This means classical macro Ohm's law is reduced to micro Ohm's law under particular boundary conditions.

The reduction process is shown in Fig. 4.7.

In this explanation, we assumed that physical symbols such as V, I, R, E, i are treated as indefinite designators. This interpretation causes the following two philosophical issues.

From our philosophical standpoint, set theory is a unified conceptual basis for describing scientific theories. Working scientists use mathematical theory based on set theory to describe scientific laws in their research areas. A scientific law is a universal law in its application area. In usual, the universality of laws is not expressed by universal sentences with universal quantifiers. The symbols of natural laws are expressed by using set-theoretical symbols such as I, V, R, i, and E. From our standpoint, these symbols are expressed as indefinite designators in set-theoretical language. The symbols are interpreted to the data structure obtained by conceptual or actual experiments. The interpretations are given in a different way depending on its experimental conditions. Thus we consider that the universality of laws comes from the variety and validity of these interpretations of the symbols.

In the philosophy of science, they consider that the universal quantifier shows the universality of natural laws in first-order language. The classical covering law model by C. G. Hempel represents this type of standpoint (Hempel, 1970). This standpoint is far from the view of working scientists.

From our philosophical standpoint with indefinite designators, the interpretation of indefinite designators gives universality of the law.

We insist that scientific laws are not expressed by universal sentences of first-order language but by closed sentences with indefinite designators of the set-theoretical language. The interpretation of indefinite designators gives universality of the law.

This means that the universality of natural law does not come from the scope of a universal sentence but from the universality of the interpretation of indefinite designators in the law. This requires a change of standard view of the philosophy of science.

4.5.3 Two Dogmas of Reductionism: Emergence and Reduction

In this section, we will focus on inter-theoretical relations in systems science.

From a philosophical point of view, the concept of reduction was restricted to simple problems such as reducing a theoretical statement to an observational statement, described by first-order predicate logic.

In systems science, the focus can be on inter-theoretical properties that are relevant to more practical cases, such as "reduction," "realization," "emergence properties (emergence)," and "causal divide." We will also discuss two contradictory dogmas of reductionism (micro-macro relations). The first dogma asserts that macro theory should be reduced to micro theory. The second dogma argues that emergent properties of macro theory may appear that cannot be explained by reduction to micro theory and that it is essential in micro-macro relations to reveal these emergent properties.

In this section, we first generalize the scheme of the reduction process discussed in the previous section: (1) Σ denotes the macroscopic phenomenological theory; (2) Γ denotes the basic theory of microscopic phenomena; (3) Γ' denotes the theory with specific boundary conditions added to Γ; (4) $\Gamma'\mathrm{ex}$ denotes the conserved expansion of Γ' by defining macroscopic indeterminates of Σ in $L[\Gamma']$. (5) $\Gamma'\mathrm{ex}$ denotes the conservative extension of Γ' obtained by defining macroscopic indefinite designators of Σ to $L[\Gamma']$. The generalized schema is shown in Fig. 4.8.

A natural interpretation is defined from $L[\Gamma]$ to $L[\Gamma']$ and $L[\Gamma'\mathrm{ex}]$. If there exists an interpretation J from $L[\Sigma]$ to $L[\Gamma'\mathrm{ex}]$, and the properties (axioms) of Σ hold in $\Gamma'\mathrm{ex}$ in the interpretation, then we say that Σ is reduced to the microscopic theory Γ, or that Σ is realized by the microscopic theory Γ.

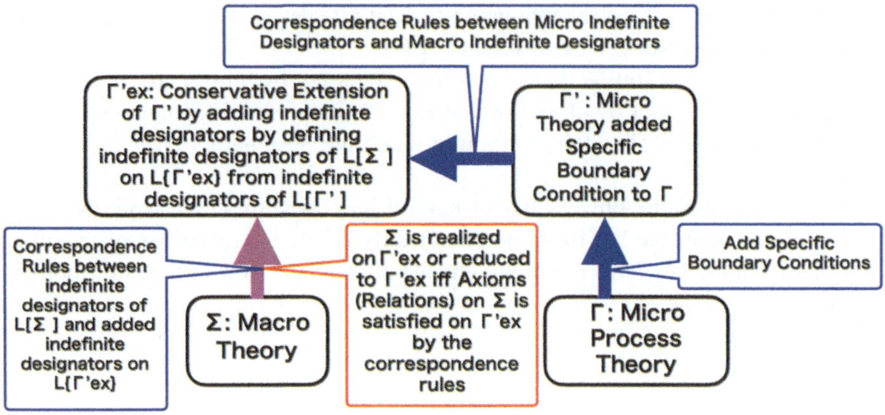

Fig. 4.8 Generalized schema of reduction and realization

We use the term reduction to give a microscopic theory basis for a macroscopic phenomenological theory. On the other hand, we use the term realization when we use this macro-micro relation to construct macroscopic structures from microscopic phenomena in the real world by adding specific boundary conditions to microscopic phenomena.

In the reduction of the macroscopic Ohm's law Σ: $V = IR$ to a microscopic theory, let $i = \kappa E$ be the microscopic theory Γ, and let Γ' be the theory with the particular boundary condition of a steady current field of cross section S, length L, and current density i. Then, by defining $I = \int Si{\cdot}n\mathrm{d}s$, $V = -\int E{\cdot}n\mathrm{d}L$, and $R = L/(S\kappa)$, Γ'ex is defined as a conservative extension of Γ'. On this Γ'ex, the indefinite designators V, I, R of the macroscopic Ohm's law Σ are given interpretations, showing that $V = IR$ holds on Γ'ex. Using this relation, we can design a resistor as an electrical component and realize it in the real world.

Electromagnetic devices have been designed as systems with a special boundary of micro electromagnetic laws. They have been realized and constructed as components in the real world. The term "realization" indicates that the reduction scheme is also used to intervene in the real world and construct a real system.

Next, we introduce the concept of the causal divide. There is no causal divide between macro and micro phenomena in the previous schema of reduction and realization. Then macro phenomena are caused and ruled by micro phenomena.

On the other hand, the following cases show that the macro phenomena is realized by micro phenomena that realize macro phenomena, but micro phenomena do not rule the macro phenomena causally.

One example is computer programs. "A Computer program" on a computer is not ruled by the Maxwell Equation on the electromagnetic field with specific boundary conditions but is realized by electromagnetic components. A program rules the function on the computer.

The other example is organizational roles. "A Role" of an organization is not reduced to an agent activity but is realized by the agent who take the role.

The first case shows that when logic gates are realized not by electromagnetic components but by other physical phenomena such as fluid dynamics, the computer will not be realized by electromagnetic components.

The second case shows computer-controlled robots might realize the role instead of human agents.

These cases show that the macroscopic functional system at the upper level is realized by some microscopic system at the lower level. However, the realization of these systems is not unique. Of course, a certain lower-level microsystem realizes the upper-level macrosystem. Thus the micro phenomena affect the macro phenomena through the realization process. Nevertheless, it will not rule the macro function. In other words, if a computer malfunctions, the programs on that computer will not work correctly either.

However, repairing the malfunction does not mean bug fixing of the computer program.

The lower-level microsystem does not govern the functions of the upper-level macrosystem; the functions of the upper-level macrosystem are described on the upper-level system like a computer program. The organizational roles are also a kind of program of the functions of the organization.

We will conclude this chapter by giving an example of a specific hierarchical realization of the "Causal Divide." It is too complex to discuss the realization process of a computer and its programs from electromagnetic phenomena. Instead, we will show a realization process of the Half Adder, which performs a one-digit calculation in binary notation from electromagnetic phenomena.

A single-digit binary calculation is mathematically formulated and causally divided from the law of Quantum electrodynamics (QED). This calculation is performed by a circuit called the Half Adder, which is a system that imposes circuit constraints on logic gates. Transistors and diodes realize the logic gates. The transistors and diodes are realized by semiconductors based on Quantum electrodynamics. This hierarchical relationship is shown in Fig. 4.9.

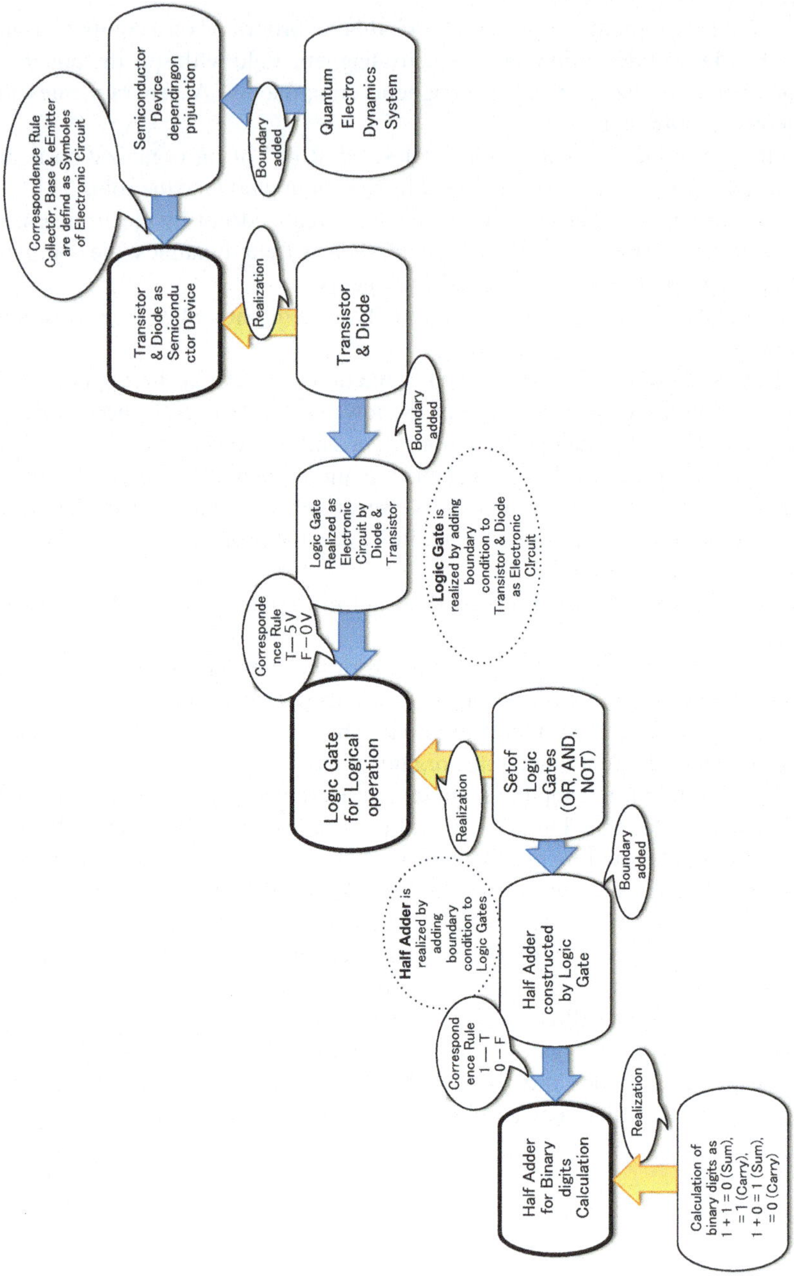

Fig. 4.9 Realization hierarchy from single-digit binary calculation to QED

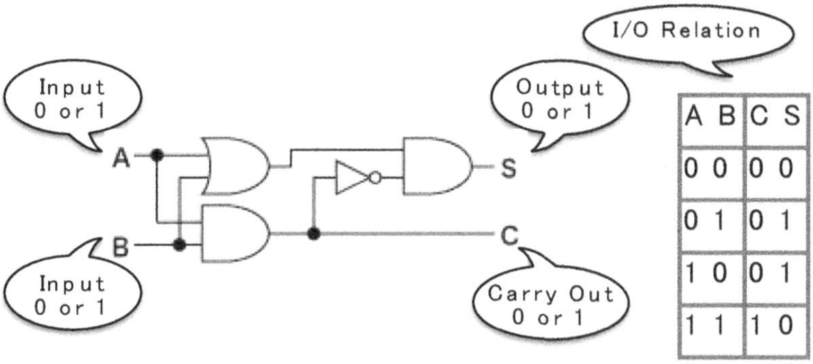

Fig. 4.10 Design for constructing a half adder from logic gates

Figure 4.10 shows a design for constructing a half adder by imposing circuit constraints on logic gates.

Figure 4.11 shows a design for constructing logic gates from transistor and diode by imposing a specific boundary condition.

Figure 4.12 shows a design of NPN transistor based on QED.

4.6 Conclusion

This chapter has discussed inter-theoretical relations by introducing indefinite designators in the semantics of the set-theoretical language. An inter-theoretical network gives an intensional meaning of indefinite designators in sentences of the theory in the network, which means a semantic holism in the network. The interpretation of indefinite designators in a structure of the ZFC Universe gives an extensional meaning in a model. The interpretation gives real-world extensional meaning if a specific observation method constructs the model from real-world data.

We have also clarified the concept of reduction and realization between macro and micro theories. The reduction process explains how macro phenomenological theory can be reduced to a micro mechanism with specific boundary conditions and symbol mapping between macro theory and micro theory. The realization process gives design how macro function can be constructed from micro mechanism with specific boundary conditions and symbol mapping between functional macrosystem and micro mechanism. The realization design can be applied to the real-world construction of macro functions.

The meta-analysis provides knowledge pragmatics for working systems scientists in natural and engineering domains and social systems (Deguchi, 2021).

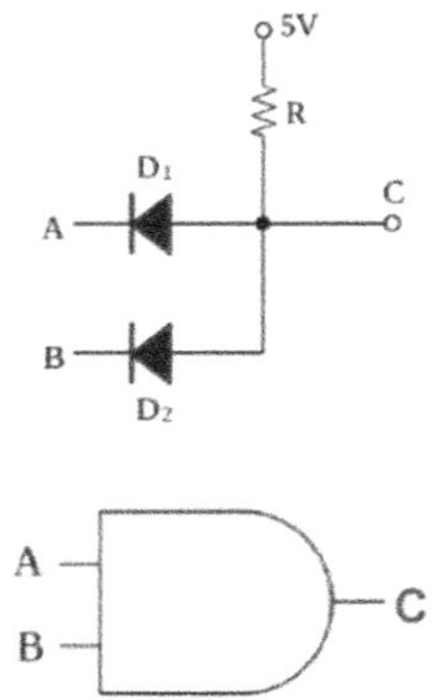

A	B	C
5V	5V	5V
5V	0V	5V
0V	5V	5V
0V	0V	0V

A	B	C
T	T	T
T	F	T
F	T	T
F	F	F

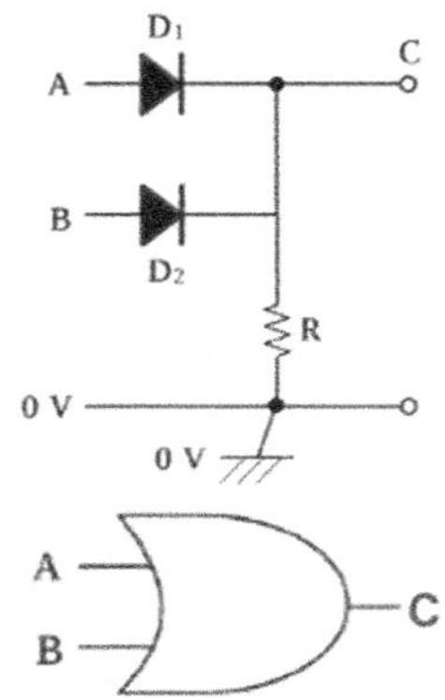

A	B	C
5V	5V	5V
5V	0V	5V
0V	5V	5V
0V	0V	0V

A	B	C
T	T	T
T	F	T
F	T	T
F	F	F

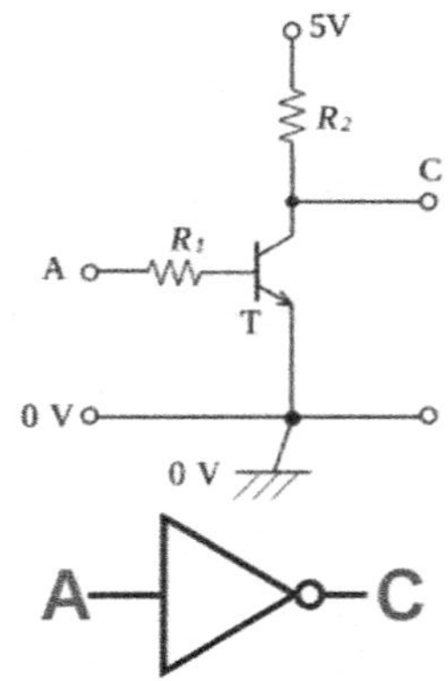

A	C
5V	0V
0V	5V

A	C
T	F
F	T

Fig. 4.11 Design for constructing logic gates from transistor and diodes

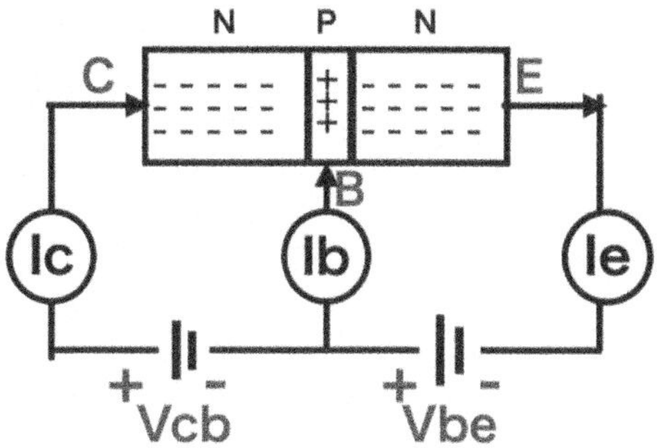

Fig. 4.12 Design of NPN transistor based on QED

References

Allwood, J., Andersson, L.-G., & Dahl, O. (1977). *Logic in linguistics (Cambridge textbooks in linguistics)*. Cambridge University Press.

Bourbaki, N. (2008). *Theory of sets (elements of mathematics)*. Springer.

Deguchi, H. (2021). Social systems as the artificial. In G. S. Metcalf, K. Kijima, & H. Deguchi (Eds.), *Handbook of systems sciences* (pp. 675–698). Springer.

Fodor, J. A., & Lepore, E. (1992). *Holism*. John Wiley & Sons.

Graetzer, G. (1979). *Universal algebra*. Springer.

Hempel, C. G. (1970). *Aspects of scientific explanation and other essays in the philosophy of science*. The Free Press.

Husserl, E. (2012). Formal and transcendental logic (English Edition, D. Cairns, trans.), Springer.

Mesarovic, M. D., & Takahara, Y. (1975). *General systems theory: Mathematical foundation*. Academic Press.

Mesarovic, M. D., & Takahara, Y. (1989). *Abstract systems theory*. Springer-Verlag.

Quine, W. V. O. (1980). *From a logical point of view: Nine logico-philosophical essays* (second revised edition). Harvard University Press.

Suppe, F. (1977). *The structure of scientific theories* (2nd ed.). University of Illinois Press.

Takeuti, G., & Zaring, W. M. (1982). *Introduction to axiomatic set theory* (2nd ed.). Springer.

Chapter 5
Beyond Logical Approach to Systems Theory

Shingo Takahashi

Abstract Logical Approach to Systems Theory (LAST) provides a "meta"-framework to describe and investigate explicitly and deeply the similarity of system models based on general and formal definitions of system models and their structures. The main theorem of LAST is F-morphism theorem. It substantially enhances the concept of "isomorphism" between system models of the same type to those of "different types" in the sense that they can be described in different types of languages. After considering the limitation of LAST by interpreting Gödel's Incompleteness Theorem, it is clarified that the adaption and the structural change of a system would be beyond the description capability of LAST. Hence some new conceptual devices and effective models such as agent, internal model, and organizational learning are required to be developed. Agent-Based Organizational Cybernetics (AOC) could be a key model to describe the organizational learning that induces the adaptation and structural change of system models especially in social systems.

Keywords System model · Structure · Logic · Language · Homomorphism · F-morphism · Internal model · Agent · Organizational learning

5.1 Introduction

This paper describes the essence of Logical Approach to Systems Theory (LAST) (Takahashi & Takahara, 1995), and the framework of agent-based organization cybernetics (AOC) for consideration for adaptation that includes internal model and organizational learning as the main concepts (Takahashi, 2006).

S. Takahashi (✉)
School of Creative Science and Engineering, Waseda University, Tokyo, Japan
e-mail: shingo@waseda.jp

K. Kijima et al. (eds.), *Systems Research I*, Translational Systems Sciences 26,
https://doi.org/10.1007/978-981-16-9937-5_5

The similarity of systems models has been a central concept in systems theory as well as systems science. Basically, from the theoretical point of view, a system model is similar to another if a homomorphism can be defined between the two system models. The homomorphism theorem is one of the typical results on how isomorphic images can be constructed in terms of homomorphism. Though the homomorphism concept should be primally considered to give a similarity relation of systems models, it should be still clarified what properties are interpreted as similar between system models in the sense of "preserving" the properties by a homomorphism from one model to another, and how the system model should be expressed to define in a totally formal way the similarity by generalizing the homomorphism concept.

LAST will give an answer to these questions on general similarity.

LAST provides a "meta"-framework to describe and investigate explicitly and deeply the similarity of system models based on general and formal definitions of system models and their structures. Systems models are described in terms of model theory. Roughly speaking, the following relationship holds (Chang & Keisler, 1973):

$$\text{model theory} = \text{universal algebra} + \text{logic}.$$

Universal algebra is appropriate or proper for describing system models, especially general system models or abstract system models (Mesarovic & Takahara, 1975, 1989), each of which is expressed as a mathematical structure. Logic generally includes as the main body language, formation rules of formulas, deduction system, and satisfaction of formulas in a model. We sometimes use category theoretical formulation as well as universal algebra, which is often effective for the theory of general system models, when we consider a class of all systems satisfying some specific properties, for example, a class of all state space representations.

The main theorem of LAST is F-morphism theorem. It substantially enhances the concept of "isomorphism" between system models of the same type to those of "different types," which means that the two models described in "different languages" can be isomorphic in terms of F-morphism. The relation of being "isomorphic" of system models is the basis of the similarity in systems science. Hence the F-morphism concept in LAST should be considered as fundamental for the similarity in systems science, as well as homomorphism as a specific case.

The logical approach also clarifies the distinction of what type of properties of systems can be described in the theory and what type of them cannot. In particular, the adaptation of systems and the structural change of a system would be beyond the description capability of LAST. We need to develop some new conceptual devices such as agent, internal model, and organizational learning. This chapter focuses on considering the adaptation of social systems and introduces the agent-based organization cybernetics (AOC) as a key model.

5.2 General Systems

In the most general sense, a system can be defined as a relation on some attributes $V_1, \ldots, V_n$, the relation which is expressed by $S \subset V_1 \times \cdots \times V_n$ (Mesarovic & Takahara, 1975, 1989). Each set V_i represents the collection of alternative ways in which the corresponding object appears in the relation that defines the system. The object is identified in terms of a property and an attribute. A general form of an input-output system model is described as $S \subset X \times Y$, where X is the set of input attributes of concerns and Y the set of output attributes of concern, respectively. Starting from this general definition of a system, we develop systems theory by introducing into the attributes some structures such as linearity, stationarity, and so on that are suitable to our interests in objects as systems. Our definition of a system model realizes this concept of a system in as a formal and general way as possible.

5.3 Logical Approach to Systems Theory: LAST

Logical Approach to Systems Theory (LAST) provides a second-order framework to describe and investigate explicitly and deeply both system models and structures of them from a model theoretic point of view (Takahashi, 1995). LAST is characterized by type-free representation, distinction between model and structure, and hierarchical structure expansion.

1. Type-free representation of system models. The representation is not only independent of any specific formalisms such as differential equations or automata, but also capable of clarifying possibly different types of system models constructed from multifaceted aspects in modeling.
2. Distinction of system models from their structures. The logical approach provides a language and formal framework to describe the properties of each system model, to define its structure and specify the class. The language is determined by the structure of a system model.
3. Hierarchical structure expansion. The relations of inclusion among classes of system models are given as hierarchical relations of the structures of system models. That is, when a class of system models is included in another, the structure of system models of the former class is obtained by expanding the other. The expanded structure inherits the antecedent.

LAST places its emphasis mainly on complex systems such as information systems or social systems as concrete instances rather than traditional topics in control theory. So complex is even a single information system that includes system models of various types coming from diversity of individual objects. The design of a complex system requires to deal with such variety of types of system models and to specify a class of system models of any type separately from the description of the system models. Thus exploring the inter-relations among system models to describe a complex system is expected as a key feature in LAST. The primary purpose of LAST is to provide an effective framework for using basic concepts in systems theory to describe complex systems. The use of model theory rather than usual set theory would make this purpose easily attainable. LAST aims at a practical device for designing complex systems as well as theoretical development on system models and structures.

5.3.1 Basic Concepts of LAST

There are at least four basic concepts to be understood in applying LAST. We will illustrate in the following sections each of the concepts in detail.

1. **System model.** As stated previously, system models are objects of study in systems theory. LAST provides a formal framework for representing a system model to reflect systems recognition of a model builder. The representation should be fully independent of the types of system models, while individual system models employed in individual systems theories have their own specified types. Thus we have to specify the language in such a way that not only system models, but their types can be described, that is, what the types of system models are should be clearly defined. LAST gives a natural and suitable way to satisfy such requirement.
2. **Structure.** Since every system in the reality is recognized only as a system model, the structure of a system is equivalent to that of a system model. If it is allowed to use the term "structure" in defining a system, we could define a system as follows: "A system is a whole entity having its own structure."

   The concept of structure has been less well-defined than that of a system and rather controversial. However, in developing a meta-theory of systems, we cannot avoid making clear the concept of structure in a formal way. In our logical approach structure of a system model will be defined by a pair of a language to describe the system model and a set of formulas to specify the behavior of the system model. This definition comprehends essential parts of other definitions of structure.
3. **Morphism.** The morphism is a conceptual basis for considering similarities between system models. The similarity between two system models is often defined by some morphism between them, more precisely, by some homomorphism. The definition of similarity by homomorphism, however, depends on a particular representation or specification such as Mealy type automata used to describe the situation. Since morphism is both practically and conceptually significant as such in systems theory, we need to develop some general morphism independent from representation types so that it gives the similarity between system models not only of the same kind of type, but of different types. The type-freeness of representation of system models in LAST enables us to construct such a general morphism between system models of possibly different types including

homomorphism as a truly special case. For example we can construct a general morphism from a finite automaton to a Petri net.

4. **Universality.** Since an aim of LAST is to develop a meta-theory concerning "inter-models," we are interested in universal properties found in a class of system models or their structures rather than in individual models as instances. So far, some universal properties significant in systems theory have been examined. Here we will concentrate on the realization problem as universality, the problem which deals with how the minimal model in a given class of a structure can be constructed from a given set of input-output pairs. The algebraic specification is one of important examples of the realization as universality.

5.3.2 System Model

5.3.2.1 Definition of System Model

A system model is a whole entity with some interactions among its elements. A direct and natural representation idea of a system model is to express it as a mathematical structure that consists of a base set with relations and functions defined on it.

Definition 5.1 [System Model] A system model $\mathfrak{M}$ is composed of:

1. A base set M;
2. A set of $\lambda(i)$-ary relations on M, $\{R_i | i \in I\}$, where λ is a function such that $I \to N^+$ (positive integers);
3. A set of $\mu(j)$-ary functions on M, $\{f_j | j \in J\}$, where μ is a function such that $J \to N$ (non-negative integers).

Here an n-ary relation or function has n arguments, written as $R(a_1, \ldots, a_n)$, or $f(a_1, \ldots, a_n)$. The function λ (or μ) means that the arity of a relation R_i (or f_j) depends on its index i (or j), written as $R_i(a_1, \ldots, a_{\lambda(i)})$ or $f_j(a_1, \ldots, a_{\mu(j)})$. Nullary functions with no arguments are called *constants*. The pair $\langle \lambda, \mu \rangle$ is called the *type* of $\mathfrak{M}$.

We write $\mathfrak{M}$ as follows:

$$\mathfrak{M} = \langle M; \{R_i | i \in I\}, \{f_j | j \in J\} \rangle.$$

We sometimes write $| \mathfrak{M} |$ to indicate the base set M.

This definition of a system model has very wide applicability. Most systems representations we are interested in can be reformulated in the above form. We illustrate below some typical and significant examples of system models.

Example 5.1 [Input-Output System Model] A simple but quite important instance of system models is an input-output system model. Although we could describe it in some different representations, the following one is natural. An input-output system model is expressed by

$$M_{I/O} = \langle X \cup Y; S, X, Y \rangle ,$$

where $S \subset X \times Y$, X is the set of inputs and Y the set of outputs.

Example 5.2 [Linear System Model] A linear system model is a system model whose input set and output set are vector spaces and whose behavior has the linear property that $S(x, y)$ and $S(x', y')$ implies $S(\alpha x + \beta x', \alpha y + \beta y')$ for any α and β in a field F over which the input and output sets are the vector spaces. The linearity of the input set is represented by the following system model: $M_X = \langle F \cup X; F, X, +, -, ^{-1}, \cdot, 0_F, 1_F, 0_X \rangle$, where X is the set of inputs, F a unary relation that is the set of scalars, 0_X the zero vector, 0_F the zero scalar, 1_F the unit element of F, $+$ a binary function representing both scalar and vector addition, $-$ a unary function representing the additive inverse, $\cdot$ a binary function representing scalar multiplication and multiplication of a vector by a scalar and $^{-1}$ a unary function representing the multiplicative inverse. The linearity of the output set is similarly defined. A linear system model is defined as the union of, $\mathfrak{M}_Y$ and the input-output system model with the linear property of behavior, where the union of two system models $\mathfrak{M}_1$ and $\mathfrak{M}_2$ is the system model whose base set, functions, and relations are respectively the unions of the corresponding sets of the two system models.

Example 5.3 [Discrete Event System Specification (DEVS) Model] The DEVS formalism provides a means of constructing simulation models and a formal representation of discrete event systems capable of mathematical manipulation just as differential equations serve this role for continuous systems (Zeigler, 1990). A DEVS model described in the DEVS formalism consists of a time base, inputs, states, outputs, and functions for determining next states and outputs given current states and inputs:

$$M_{\mathrm{DEVS}} = \langle X \cup S \cup Y \cup R \cup \{\infty\}; X, S, Y, \delta_{\mathrm{int}}, \delta_{\mathrm{ext}}, \lambda, t_{\mathrm{a}}, Q, T \rangle ,$$

where X is a set of external event types, S a set of sequential state, Y a set of external event types generated as outputs, T the time base, t_{a} the time advance function from S to the non-negative reals with infinity:$t_{\mathrm{a}} : S \rightarrow R^{+}_{0,\infty}$, Q the total state set defined by $Q = \{(s, e) | \, s \in S, 0 \leq e \leq t_{\mathrm{a}}(s)\}$, δ_{int} the internal transition function: $\delta_{\mathrm{int}} : S \rightarrow S$, δ_{ext} the external transition function: $\delta_{\mathrm{ext}} : Q \times X \rightarrow S$, and λ the output function:$\lambda : Q \rightarrow Y$.

Some essential behavior of a system specified by DEVS, such as the property of the output function that generates an external output just before an internal transition takes place, should be considered to be included implicitly in $\mathfrak{M}_{\text{DEVS}}$.

5.3.2.2 Language for Describing Systems Properties

The systems properties are the properties possessed by a system model such as linearity, stationarity, or causality. To investigate properties of these systems properties, which can be called the meta-treatment of system models, we introduce a formal language to describe systems properties. The use of the formal language characterizes LAST. In this section we give only the formal framework of the language.

Definition 5.2 [Language for a System Model] The language for a system model $\mathfrak{M}$ consists of:

1. $\lambda(i)$-ary predicate letters $\mathbf{R}_i$ for each $i \in I$, where λ is a function such that $I \to N^+$ (positive integers);
2. $\mu(j)$-ary function symbols $\mathbf{f}_j$ for each $j \in J$, where μ is a function such that $J \to N$ (non-negative integers).

We write $\mathfrak{L}(\mathfrak{M})$ also as $\mathfrak{L}(\mathfrak{M}) = \left\langle \{\mathbf{R}_i | i \in I\}, \{\mathbf{f}_j | j \in J\} \right\rangle$.

$\langle \lambda, \mu \rangle$ is also said to be the type of $\mathfrak{L}(\mathfrak{M})$. There is the one-to-one correspondence, denoted by *Cor*, between the boldface symbols in $\mathfrak{L}(\mathfrak{M})$ and the light face symbols for the relations and functions in $\mathfrak{M}$, i.e., $Cor(\mathbf{R}_i) = R_i$ for each $i \in I$ and $Cor(\mathbf{f}_j) = f_j$ for each $j \in J$. Then $\mathfrak{M}$ is said to be a *realization* of the language $\mathfrak{L}(\mathfrak{M})$ or *model* for $\mathfrak{L}(\mathfrak{M})$. The languages for two system models of the same type are, up to alphabetic variants, identical. Therefore every system model of the same type as a system model $\mathfrak{M}$ is a realization of the language $\mathfrak{L}(\mathfrak{M})$.

For example, an input-output system model,

$$\mathfrak{M}_{I/O} = \langle X \cup Y; S, X, Y \rangle$$

and all the system models of the same type as this system model are realizations of the language,

$$\mathfrak{L}\left(\mathfrak{M}_{I/O}\right) = \langle \mathbf{S}, \mathbf{X}, \mathbf{Y} \rangle .$$

We should notice that for language we customarily use boldface symbols with the same alphabets as a system model, for example, S and $\mathbf{S}$, so as not to confuse language with system models. This usage is only for the sake of convenience. However, we should notice that a language, e.g., $\mathfrak{L}\left(\mathfrak{M}_{I/O}\right)$, is purely syntactic construct. Other models than $\mathfrak{M}_{I/O}$, e.g.,$\mathfrak{M}' = \langle Z; T, V, W \rangle$, can be also realizations of $\mathfrak{L}\left(\mathfrak{M}_{I/O}\right)$, even if they have no property of an input-output system model. The desirable properties that every input-output system model should have are specified not only as a language but as a structure of the system model.

To describe systems properties, we need "grammar" that distinguishes "right sentences" from wrong sentences. In our logical approach, we assume that the properties of a system can be expressed as first-order sentences in first-order language that plays the role of the "grammar." The first-order language consists of the primitive symbols such as variables, logical connectives, quantifiers, identity symbols, parentheses and comma with the language $\mathfrak{L}(\mathfrak{M})$ for a system model, the formation rules of the terms, the atomic formulas and the well-formed formulas. The set of terms of the language $\mathfrak{L}(\mathfrak{M})$, denoted by Term $(\mathfrak{L}(\mathfrak{M}))$, is recursively defined from the language $\mathfrak{L}(\mathfrak{M})$. Similarly the set of atomic formulas, denoted by Atom $(\mathfrak{L}(\mathfrak{M}))$ and that of well-formed formulas, denoted by Form $(\mathfrak{L}(\mathfrak{M}))$, are recursively defined from the language.

The logical connectives and universal quantifier as primitive symbols have no proper meanings such as "and," "not," and "for all." These intended meanings are realized only when these symbols are interpreted in a specific system model. This realization is called satisfaction. We will usually use other symbols, say $\mathbf{x}$, $\mathbf{y}$, $\mathbf{z}$, as individual variables. A term that has no variable is called a closed term. In first-order logic some abbreviations such as $\exists \mathbf{v}$, $\phi \vee \psi$, $\phi \rightarrow \psi$, and so on will be defined in the standard manner. Although the abbreviations actually intend to have the meanings of "for some (or there exist)," "or," and "imply" respectively, these meanings are only realized in a system model. Other logical concepts such as the *scope* of the quantifier, a bound variable, a free variable, and so on are introduced in the language, which can be found in the standard textbook of logic. A formula $\phi \in$ Form $(\mathfrak{L}(\mathfrak{M}))$ is said to be a *sentence* of $\mathfrak{L}(\mathfrak{M})$ if ϕ has no free variables. Sent $(\mathfrak{L}(\mathfrak{M}))$ denotes the set of sentences.

In our logical approach every *systems property* of an individual system model is expressed by a sentence. For example, an input-output system model $\mathfrak{M}_{I/O}$ has a basic systems property: "every element of the system is a pair of an input and an output." This property can be expressed by the following sentence: $(\forall \mathbf{xy})(\mathbf{S}(\mathbf{x}, \mathbf{y}) \rightarrow \mathbf{X}(\mathbf{x}) \vee \mathbf{Y}(\mathbf{y}))$. In ordinary mathematical notations, i.e., in set-theoretical language, this sentence means that $S \subset X \times Y$. As another example, an input-output system of function-type (Mesarovic & Takahara, 1989) can be expressed as: $(\forall \mathbf{x} \in \mathbf{X} \rightarrow (\exists ! \ \mathbf{y} \in \mathbf{Y})\mathbf{S}(\mathbf{x}, \mathbf{y}))$, where the notation $\exists ! \ \mathbf{x}\phi(\mathbf{x})$ is the abbreviation of the sentence that means "there uniquely exists $\mathbf{x}$ such that $\phi(\mathbf{x})$."

We have noticed that a systems property is expressed by a sentence. Conversely, a sentence should be interpreted as a systems property in a system model so that the sentence obtains a concrete meaning in the system model.

Let us consider a system model $\mathfrak{M}_S = \langle \{a, b, c\}; S, X, Y \rangle$, where $S = \{(a, b), (a, c), (b, c)\}$ and $X = Y = \{a, b\}$. Is a formula $\mathbf{S}(\mathbf{x}, \mathbf{y}) \rightarrow \mathbf{X}(\mathbf{x}) \vee \mathbf{Y}(\mathbf{y})$ true in this system model? If we assign a and b to the variables $\mathbf{x}$ and $\mathbf{y}$ respectively, the formula is true in that model. However if c to $\mathbf{y}$, then it is not true. To judge the truth of a formula containing some free variables, we need to assign an element of the base set of a system model to each variable. As will be stated later, since a sentence has no free variable, we can judge its truth without depending on the assignment of variables.

For example, a sentence $(\forall \mathbf{xy})(\mathbf{S}(\mathbf{x},\mathbf{y}) \rightarrow \mathbf{X}(\mathbf{x}) \vee \mathbf{Y}(\mathbf{y}))$ is not true in the system model $\mathfrak{M}_S$, so this model is not an input-output system model. An assignment to each variable is defined by an assignment function. Given a system model $\mathfrak{M}$ with a base set M, an *assignment function* ρ (or briefly *assignment*) is a function of the set V of variables to M. Then for a given assignment, sentences in terms of the language of the system model are interpreted into the system model. This interpretation is defined as the *denotation* of a term in $\mathfrak{L}(\mathfrak{M})$. Each variable $\mathbf{x}$ is replaced by the element $\rho(\mathbf{x}) \in M$ and each function symbol is replaced by the corresponding function. A denotation with respect to a given assignment can be regarded as a function of terms to the base set of a system model.

The concept that a systems property holds in a system model is defined as the *satisfaction* of formulas.

Definition 5.3 [Satisfaction] A formula ϕ holds in $\mathfrak{M}$ with an assignment function ρ, or ρ satisfies ϕ in $\mathfrak{M}$, written $\mathfrak{M} \models \phi\,[\rho]$, is defined recursively:

1. $\mathfrak{M} \models \mathbf{R_i}\left(t_1, \ldots, t_{\lambda(i)}\right)$ if and only if $\left\langle t_1^d\,[\rho], \ldots, t_{\lambda(i)}^d\,[\rho]\right\rangle \in R_i$;
2. $\mathfrak{M} \models \neg\phi\,[\rho]$ if and only if it is not the case that $\mathfrak{M} \models \phi\,[\rho]$;
3. $\mathfrak{M} \models \phi_1 \wedge \phi_2\,[\rho]$ if and only if $\mathfrak{M} \models \phi_1\,[\rho]$ and $\mathfrak{M} \models \phi_2\,[\rho]$;
4. $\mathfrak{M} \models \forall \mathbf{x}\phi\,[\rho]$ if and only if $\models \phi[\rho(y/\mathbf{x})]$ for any $y \in M$.

5.3.3 Structure

The structure of a system model characterizes the system model in the sense that the structure determines to which class of systems the system model pertain. In this sense if a system model is expressed by a collection of some differential equations, we can say that the matrices of the coefficients of the differential equations give a structure of the system model. However, from systems viewpoints, a class of system models should be specified not by the form of differential equations, but by a set of systems properties. Hence the structure of a system model should have at least the following features.

First, the structure of a system model generates its properties or behavior to be recognized.

Second, the representation of structure is based on a hierarchical construction. For example, the structure of an input-output linear system model is "hierarchically" constructed from both a linear structure and an input-output structure, in the sense that the input-output linear structure explicitly "inherits" the properties from the linear and input-output structure.

Third, the structure distinguishes the properties of the class of system models satisfying it from those of an individual system model in the class.

One way to fulfill the above requirements is to adopt a "language" that expresses systems properties, and to represent the structure as "axioms." This means that we should abstract basic properties from a class of system models as axioms that are common characteristics of the class. Thus the structure of a "family system" in the previous section is abstracted from concrete family models. For example, we can abstract some axioms such that every father is a male, every mother is a female, father and mother are married, all brothers have the same father and mother, and so on. The language such as "father," "male," "every," "is-a," etc., and some "grammar" to make legal sentences should be chosen before axioms are described. Then the axioms are expressed by some sentences in that language. We should notice that this example of the structure of family does not include all families at all; a family that has brothers whose mothers are different is not included.

A language and axioms are chosen from a systems viewpoint that reflects our current interest. In this sense the structure of a system model expresses fundamental interactions we recognize as the system model does. Thus a modeling process contains as its essential part some stages of specifying language and constructing axioms. Consequently the structure of a system model is defined as a pair, $(\mathfrak{L}; \Sigma)$, of language $\mathfrak{L}$ to define the system model and a set of axioms Σ to describe the class to which the system model pertains. In LAST every structure is defined in a formal language such as first-order language. Use of other formal languages than first-order is not restricted in LAST. We notice that there are some advantages and disadvantages of the use of first-order language.

The formal description of a structure of a system model has some technically outstanding advantages as well as conceptual ones. It enables us to point out what a systems property of a given system model is, and to distinguish the system properties from system models that "satisfy" the properties. This relation is provided as satisfaction relation that is one of the main characteristics of LAST: type-free representation. Thus we can construct and specify a class of system models without depending on the concrete descriptions of individual system models.

The formal definition of the structure of a system is defined below.

Definition 5.4 [Structure of a System] Let $\mathfrak{M}$ be a system model, $\mathfrak{L}(\mathfrak{M})$ the language for $\mathfrak{M}$, and Σ a set of sentences of $\mathfrak{L}(\mathfrak{M})$, where $\mathfrak{M} \models \Sigma$. Then the *structure* of a system as a prototype of the system model $\mathfrak{M}$ is defined by $(\mathfrak{L}(\mathfrak{M}); \Sigma)$.

A given system model necessarily determines $\mathfrak{L}(\mathfrak{M})$, unique up to alphabetic invariants. We should notice that $\mathfrak{L}(\mathfrak{M})$ is a collection of "symbols," therefore the role of $\mathfrak{L}(\mathfrak{M})$ in the systems recognition is to point out the names and types of the relations that are identified in the system we recognize.

On the other hand, Σ provides the rules how elements in a system model interact. Therefore the properties of a system implied by the structure of the system are expressed as the formulas derived from Σ; $T(\Sigma) = \{\phi \in \text{Sent}(\mathfrak{L}(\mathfrak{M})) \mid \Sigma \vdash \phi\}$ is the whole of the properties characterized by the structure of the system, $(\mathfrak{L}(\mathfrak{M}); \Sigma)$. If Σ is complete, the properties satisfied by a system model having the structure $(\mathfrak{L}(\mathfrak{M}); \Sigma)$ accord with the properties of a system implied by

$(\mathfrak{L}(\mathfrak{M})\,;\,\Sigma)$); that is, let $\mathrm{Th}(\mathfrak{M}) = \{\phi \in \mathrm{Sent}(\mathfrak{L}(\mathfrak{M}))\,|\mathfrak{M} \models \phi, \mathfrak{M} \models \Sigma\}$, then $\mathrm{Th}(\mathfrak{M}) = T(\Sigma)$. Notice that it follows from the definition that $(\mathfrak{L}(\mathfrak{M})\,;\,\Sigma)$ cannot be uniquely determined for one system model $\mathfrak{M}$ since we can take another Σ as axioms for which $\mathfrak{M}$ is a model. This means that there may be plenty of system models satisfying a given structure $(\mathfrak{L}(\mathfrak{M})\,;\,\Sigma)$. For example, many models satisfy the Peano's Axioms well known as a structure of the natural number. They are not necessarily isomorphic to the natural numbers (Chang & Keisler, 1973).

As an example we define below the structure of input-output system.

Definition 5.5 [The Structure of Input-Output System] The structure of input-output system model is defined by $(\mathfrak{L}_{I/O};\Sigma_{I/O})$:

$$\mathfrak{L}_{I/O} = \{\mathbf{X}, \mathbf{Y}, \mathbf{S}\},$$

where

$\mathbf{X}, \mathbf{Y}$: unary relation symbols,
$\mathbf{S}$: a binary relation symbol;

$$\Sigma_{I/O} = \{\phi_{I/O}\},$$

where $\phi_{I/O} \equiv (\forall \mathbf{xy})(\mathbf{S}(\mathbf{x}, \mathbf{y}) \to \mathbf{X}(\mathbf{x}) \wedge \mathbf{Y}(\mathbf{y}))$.

An input-output system model $M_{I/O} = \langle X \cup Y; S, X, Y\rangle$ is a realization of $\mathfrak{L}_{I/O}$ and model for $\Sigma_{I/O}$.

5.3.4 *Morphism*

One of the most important purposes of systems science is to investigate the similarity between system models. The similarity can be divided into two types: structural similarity and behavioral similarity. So far, fixing the type of models, we have studied structural similarity in systems theory using modeling morphisms defined especially between input-output system models. However, as will be made clear in the subsequent discussions, these modeling morphisms are defined not in a general way in which we can deal with structural similarity between any system models, but in a specific way based only on homomorphisms. This section is devoted to the development of a general theory of structural similarity between system models.

5.3.4.1 Morphisms for Models of the Same Type

In this section we investigate morphisms between system models of the same type, following the three cases mentioned in the previous section.

Preservation of Generator: Homomorphism

A morphism between system models of the same type is usually given by a homomorphism, which preserves only atomic formulas as the generators of the language.

Definition 5.6 [Homomorphism] Let $\mathfrak{M}_1 = \langle M_1; \{R_i{}^1 | i \in I\}, \{f_j{}^1 | j \in J\}\rangle$ and $\mathfrak{M}_2 = \langle M_2; \{R_i{}^2 | i \in I\}, \{f_j{}^2 | j \in J\}\rangle$. Notice that $\mathfrak{M}_1$ and $\mathfrak{M}_2$ are of the same type. A function $h : M_1 \to M_2$ is called a *homomorphism* of $\mathfrak{M}_1$ to $\mathfrak{M}_2$ if for any $i \in I$, $j \in J$, $a_1, \ldots, a_{\lambda(i)}, a_1, \ldots, a_{\mu(j)}, a \in M_1$, $(a_1, \ldots, a_{\lambda(i)}) \in R_i^1$ implies $(h(a_1), \ldots, h(a_{\lambda(i)})) \in R_i^2$ and $h\left(f_j^1(a_1, \ldots, a_{\mu(j)})\right) = f_j^2(h(a_1), \ldots, h(a_{\mu(j)}))$.

A bijective (i.e., one-to-one and onto) homomorphism is called an *isomorphism*.

From Definition, we can see that a homomorphism preserves only the atomic formulas, which is viewed as the generators of the language for a system model. In systems theory, the concept of a homomorphism is defined as a modeling morphism between input-output system models.

Definition 5.7 [Modeling Morphism (Mesarovic & Takahara, 1975, 1989)] Let $S \subset X \times Y$ and $S' \subset X' \times Y'$ be input-output system models. Let $h_x : X \to X'$ and $h_y : Y \to Y'$ be functions. $h = (h_x, h_y) : S \to S'$ is called a *modeling morphism of S to S′* if for any $(x, y) \in X \times Y$, $(x, y) \in S$ implies $(h_x(x), h_y(y)) \in S'$.

For example, let us consider input-output system models $\mathfrak{M} = \langle X \cup Y; S, X, Y\rangle$ and $\mathfrak{M}' = \langle X' \cup Y'; S', X', Y'\rangle$, where S, S': binary relations on $X \cup Y$ and X, X', Y, Y': unary relations on $X \cup Y$. Suppose that $\mathfrak{M}$ and $\mathfrak{M}'$ satisfy $\mathbf{S}(\mathbf{X}, \mathbf{Y}) \to \mathbf{X}(\mathbf{X}) \wedge \mathbf{Y}(\mathbf{Y})$ and $\mathbf{S}'(\mathbf{X}, \mathbf{Y}) \to \mathbf{X}'(\mathbf{X}) \wedge \mathbf{Y}'(\mathbf{Y})$, respectively. Then a homomorphism h of $\mathfrak{M}$ to $\mathfrak{M}'$ is regarded as a modeling morphism of S to S'. Notice that from the definition of a homomorphism, $h(x) \in X'$ for any $x \in X$ and $h(y) \in Y'$ for any $y \in Y$.

Preservation of Σ: Σ-Homomorphism

Next we consider homomorphisms preserving axioms Σ. Recall that the axioms Σ provide the structure of a system.

Grätzer defined such homomorphisms as Σ-homomorphisms (Grätzer, 1979). By a Σ-homomorphism the axioms Σ are preserved in a homomorphic image. We formulate a Σ-homomorphism directly based on this idea. This definition is different from Grätzer's original definition that uses the concept of $\Phi - l$ inverse.

Let h be a homomorphism of $\mathfrak{M}_1$ to $\mathfrak{M}_2$. *The homomorphic image of h* in $\mathfrak{M}_2$ is a submodel of $\mathfrak{M}_2$ whose domain is $h(\mathfrak{M}_1)$. We write $h(\mathfrak{M}_1)$ to indicate the homomorphic image of h in $\mathfrak{M}_2$ as follows.

$$h(\mathfrak{M}_1) = \left\langle h(M_1); \left\{R_i^2 \cap h(M_1)^{\lambda(i)} | i \in I\right\}, \left\{f_j^2 \parallel h(M_1)^{\mu(j)} | j \in J\right\}\right\rangle,$$

where $f_j^2 \parallel h(M_1)^{\mu(j)}$ denotes the restriction of f_j^2 to $h(M_1)^{\mu(j)}$.

By the property of a homomorphism, $f_j^2 \parallel h(M_1)^{\mu(j)}$ is well-defined.

We define a Σ-homomorphism as a homomorphism whose homomorphic image preserves Σ.

Definition 5.8 [Σ-Homomorphism] Let $\mathfrak{M}_1$ and $\mathfrak{M}_2$ be system models of the same type, and $\mathfrak{M}_1 \models \Sigma$ and $\mathfrak{M}_2 \models \Sigma$. A homomorphism h of $\mathfrak{M}_1$ to $\mathfrak{M}_2$ is called a Σ-*homomorphism* of $\mathfrak{M}_1$ to $\mathfrak{M}_2$, if $h(\mathfrak{M}_1) \models \Sigma$.

Our definition of Σ-homomorphism is slightly weaker than Grätzer's definition using the concept of $\Phi - l$ inverse. His definition requires that any "inverse" elements should be preserved. A homomorphism that has the $\Phi - l$ inverses is defined as a strong Σ-homomorphism. The axioms Σ are preserved in the homomorphic image on h in $\mathfrak{M}_2$ by a strong Σ-homomorphism h. So the homomorphic image of a strong Σ-homomorphism h, $h(\mathfrak{M}_1)$, is a model of Σ, i.e., $h(\mathfrak{M}_1) \models \Sigma$.

Preservation of Th ($\mathfrak{M}$): S-Homomorphism

That two system models are isomorphic or of the same structure implies that in a sense the properties of the two system models are equivalent. A usual homomorphism preserves the primitive properties, i.e., the generators. In this section we will define a homomorphism as an *S-homomorphism* that preserves all sentences satisfied in a system model (Th ($\mathfrak{M}$)). Furthermore we will show that an induced homomorphism is an S-homomorphism, which is well known as the homomorphism theorem. From this theorem we can see that every morphism for the structural similarity should be an S-homomorphism.

Definition 5.9 [S-Homomorphism] Let $\mathfrak{M}_1$ and $\mathfrak{M}_2$ be system models of the same type, and $h : \mathfrak{M}_1 \to \mathfrak{M}_2$ a homomorphism of $\mathfrak{M}_1$ to $\mathfrak{M}_2$. Then h is called an *S-homomorphism* of $\mathfrak{M}_1$ to $\mathfrak{M}_2$ if for any sentence ϕ of $\mathfrak{L}(\mathfrak{M}_1)$

$$\mathfrak{M}_1 \models \phi \quad \text{if and only if } h(\mathfrak{M}_1) \models \phi.$$

From the definition we can immediately see that an S-homomorphism is a Σ-homomorphism. We should notice that if every sentence that holds in $\mathfrak{M}_1$ holds in $h(\mathfrak{M}_1)$ as well, then h is already an S-homomorphism. Indeed if a sentence ϕ holds in $h(\mathfrak{M}_1)$ and does not hold in $\mathfrak{M}_1$, then $\neg\phi$ holds in $\mathfrak{M}_1$. Thus, by the above condition, $\neg\phi$ holds in $h(\mathfrak{M}_1)$, which is a contradiction.

Let $h : \mathfrak{M}_1 \to \mathfrak{M}_2$ be a homomorphism of $\mathfrak{M}_1$ to $\mathfrak{M}_2$. Then we define *the quotient system model with respect to h*, written by $\mathfrak{M}_1/h$:

$$\mathfrak{M}_1/h = \left\langle M_1/h; \left\{ R_i^1/h | i \in I \right\}, \left\{ f_j^1/h | j \in J \right\} \right\rangle,$$

where $\mathfrak{M}_1 = \langle M_1; \{ {R_i}^1 | i \in I \}, \{ {f_j}^1 | j \in J \} \rangle$ and $\mathfrak{M}_2 = \langle M_2; \{ {R_i}^2 | i \in I \}, \{ {f_j}^2 | j \in J \} \rangle$; M_1/h is the partitioned set of M_1 by the equivalence relation defined by:

$$a \equiv b \quad \text{if and only if} \quad h(a) = h(b) \quad \text{for any} \quad a, b \in M_1,$$

$s\left([a_1], \ldots, \left[a_{\lambda(i)}\right]\right) \in R_i^1/h$ if and only if $\left(h(a_1), \ldots, h\left(a_{\lambda(i)}\right)\right) \in R_i^2$, $f_j^1/h\left([a_1], \ldots, \left[a_{\mu(j)}\right]\right) = \left[f_j^1\left(a_1, \ldots, a_{\mu(j)}\right)\right]$.

$[a_i]$ denotes the equivalence class of a_i.

The quotient system model, $\mathfrak{M}_1/h$, is obviously well-defined. The well-known homomorphism theorem can be considered as a theorem by which an S-homomorphism, $h^\#$, is induced.

Theorem 5.1 [Homomorphism Theorem] *Let $\mathfrak{M}_1$ and $\mathfrak{M}_2$ be system models of the same type, and h a homomorphism of $\mathfrak{M}_1$ onto $\mathfrak{M}_2$. Then a map*

$$h^\# : \mathfrak{M}_1/h \to \mathfrak{M}_2$$

is an S-homomorphism, where $h^\#$ is defined by:

$$h^\#([a]) = h(a) \quad \text{for any} \quad [a] \in M_1/h$$

$h^\#$ is called the induced homomorphism of h.

A difference between the well-known homomorphism theorem in the usual form in algebra and this theorem is that theorem shows that the induced isomorphism $h^\#$ is an S-homomorphism preserving sentences. One of the reasons why the concept of a homomorphism is important is because an S-homomorphism $h^\#$ can be constructed from a homomorphism h.

5.3.4.2 Morphisms for Models of Different Types

In this section we consider morphisms between system models of different types. Since a homomorphism as seen in the previous section can be defined only for system models of the same type, the concept of a homomorphism is not applicable to the class of system models of different types. We, therefore, introduce a new morphism, called F-morphism, which is a generalization of homomorphism and can be applied to the class of system models of different types as well as of the same type.

In this section we also consider the three cases for the preservation of properties—generators, Σ, and Th ($\mathfrak{M}$). We will show F-morphism theorem as a theorem corresponding to the homomorphism theorem.

Preservation of Generator: F-Morphism

First we define a basic interpretation function and a basic morphism. An F-morphism is defined recursively by using these functions.

Definition 5.10 [Basic Interpretation Function] Let $\mathfrak{M}_1 = \langle M_1; \{R_i{}^1 | i \in I_1\}, \{f_j{}^1 | j \in J_1\}\rangle$ and $\mathfrak{M}_2 = \langle M_2; \{R_i{}^2 | i \in I_2\}, \{f_j{}^2 | j \in J_2\}\rangle$ be system models of possibly different types.

Then a function Bas of $\mathfrak{L}(\mathfrak{M}_1)$ to the set of formulas of $\mathfrak{L}(\mathfrak{M}_2)$ is said to be a *basic interpretation function of* $\mathfrak{L}(\mathfrak{M}_1)$ to $\mathfrak{L}(\mathfrak{M}_2)$ if the following conditions are satisfied.

1. For every relation symbol $\mathbf{R}_i{}^1 \in \mathfrak{L}(\mathfrak{M}_1)$, Bas($\mathbf{R}_i{}^1$) is a $\lambda_1(i)$-ary formula of $\mathfrak{L}(\mathfrak{M}_1)$;
2. For every function symbol $\mathbf{f}_i{}^1 \in \mathfrak{L}(\mathfrak{M}_1)$, Bas($\mathbf{f}_i{}^1$) is a $(\mu_1(j) + 1)$-ary formula of $\mathfrak{L}(\mathfrak{M}_2)$.

A basic interpretation function associates a formula of the second system model with each symbol of the language of the first one. The association is intended to give "interpretation" of the first system model to the second one. The basic interpretation function works as a meaningful interpretation only when a *basic morphism* with it is defined as follows.

Definition 5.11 [Basic Morphism] Let $\mathfrak{M}_1$ and $\mathfrak{M}_2$ be as above and Bas be a basic interpretation function of $\mathfrak{L}(\mathfrak{M}_1)$ to $\mathfrak{L}(\mathfrak{M}_2)$.

A function I_O of M_1 to M_2 is said to be a *basic morphism* of $\mathfrak{M}_1$ to $\mathfrak{M}_2$ with Bas if the following conditions are satisfied.

1. For every relation symbol $\mathbf{R}_\mathbf{i}^\mathbf{1} \in \mathfrak{L}(\mathfrak{M}_1)$ and every assignment ρ, if $\mathfrak{M}_1 \models \mathbf{R}_\mathbf{i}^\mathbf{1}\left(\mathbf{x}_1, \ldots, \mathbf{x}_{\lambda_1(\mathbf{i})}\right)[\rho]$, then $\mathfrak{M}_2 \models \mathrm{Bas}\left(\mathbf{R}_\mathbf{i}^\mathbf{1}\right)\left(\mathbf{x}_1, \ldots, \mathbf{x}_{\lambda_1(\mathbf{i})}\right)[I_O \circ \rho]$, where $I_O \circ \rho$ denotes the composition of I_O andρ;
2. For every function symbol $\mathbf{f}_\mathbf{j}^\mathbf{1} \in \mathfrak{L}(\mathfrak{M}_1)$ and every assignment ρ, if $\mathfrak{M}_1 \models \left(\mathbf{f}_\mathbf{j}^\mathbf{1}\left(\mathbf{x}_1, \ldots, \mathbf{x}_{\mu_1(\mathbf{j})}\right) = \mathbf{x}_{\mu_1(\mathbf{j})+1}\right)[\rho]$, then $\mathfrak{M}_2 \models \mathrm{Bas}\left(\mathbf{f}_\mathbf{j}^\mathbf{1}\right)\left(\mathbf{x}_1, \ldots, \mathbf{x}_{\mu_1(\mathbf{j})+1}\right)$ $[I_O \circ \rho]$, and satisfies the following condition expressing that Bas $\left(\mathbf{f}_\mathbf{j}^\mathbf{1}\right)$ is a function: $\mathfrak{M}_2 \models \left(\forall \mathbf{x}_1 \ldots \mathbf{x}_{\mu_1(\mathbf{j})}\right)\left(\exists \mathbf{x}_{\mu_1(\mathbf{j})+1}\right)\left(\forall \mathbf{y}_{\mu_1(\mathbf{j})+1}\right)\left(\mathrm{Bas}\left(\mathbf{f}_\mathbf{j}^\mathbf{1}\right)\right.$ $\left.\left(\mathbf{x}_1, \ldots, \mathbf{x}_{\mu_1(\mathbf{j})}, \mathbf{y}_{\mu_1(\mathbf{j})+1}\right) \leftrightarrow \mathbf{x}_{\mu_1(\mathbf{j})+1} = \mathbf{y}_{\mu_1(\mathbf{j})+1}\right)$.

Bas $\left(\mathbf{R}_\mathbf{i}^\mathbf{1}\right)$ and Bas $\left(\mathbf{f}_\mathbf{j}^\mathbf{1}\right)$ are called *basic interpretations* of $\mathbf{R}_\mathbf{i}^\mathbf{1}$ and $\mathbf{f}_\mathbf{j}^\mathbf{1}$. The identity $=$ is interpreted as the identity of $\mathfrak{L}(\mathfrak{M}_2)$, that is, Bas $\left(=_{\mathfrak{L}(\mathfrak{M}_1)}\right) \equiv =_{\mathfrak{L}(\mathfrak{M}_2)}$.

Definition 5.12 [F-morphism] Let $\mathfrak{M}_1$ and $\mathfrak{M}_2$ be as in Definition of Basic Morphism]. An *F-morphism*, $I : \mathfrak{M}_1 \to \mathfrak{M}_2$, is a pair of functions $\langle I_O, I_F \rangle$, where I_O is a basic morphism of with Bas and I_F is a function of the set of formulas of $\mathfrak{M}_1$ to the set of the formulas of $\mathfrak{M}_2$, which is defined as follows.

For any formula Φ of $\mathfrak{M}_1$

1. If Φ is an atomic formula of the form $\mathbf{f_j}(u_1, \ldots, u_{\mu(j)}) = \mathbf{x}$ or $\mathbf{x} = \mathbf{f_j}(u_1, \ldots, u_{\mu(j)})$, then

$$I_F(\Phi) = \begin{cases} \text{Bas}\left(\mathbf{f_j}\right)\left(u_1, \ldots, u_{\mu(j)}, \mathbf{x}\right), & \text{if } T(\Phi) \text{ is the empty set} \\ \left(\exists \mathbf{x_{k_1}} \ldots \mathbf{x_{k_m}}\right) & \\ \times \left(\text{Bas}\left(\mathbf{f_j}\right)\left(\mathbf{x_1}, \ldots, \mathbf{x_{\mu(j)}}, \mathbf{x}\right)\right. & \\ \times \wedge \left(\wedge \left(I_F\left(\mathbf{x_{k_i}} = u_{k_i}\right) | u_{k_i}\right.\right. & \\ \left.\left.\left.\in T(\Phi)\right)\right)\right), & \text{if } T(\Phi) = \left\{u_{k_1}, \ldots, u_{k_m}\right\} \end{cases},$$

where every $\mathbf{x_{k_p}}$ is a variable not occurring in Φ, and $\mathbf{x_i}$ is u_i for $u_i \notin T(\Phi)$.
2. If Φ is an atomic formula $\mathbf{P}(t_1, \ldots, t_n)$ other than of the form in (1), then

$$I_F(\Phi) = \begin{cases} \text{Bas}(\mathbf{P})(u_1, \ldots, u_n), & \text{if } T(\Phi) \text{ is the empty set} \\ \left(\exists \mathbf{x_{k_1}} \ldots \mathbf{x_{k_m}}\right) & \\ \times (\text{Bas}(\mathbf{P})(\mathbf{x_1}, \ldots, \mathbf{x}_n) & \\ \wedge \left(\wedge \left(I_F\left(\mathbf{x_{k_i}} = u_{k_i}\right) | u_{k_i}\right.\right. & \\ \left.\left.\left.\in T(\Phi)\right)\right)\right), & \text{if } T(\Phi) = \left\{u_{k_1}, \ldots, u_{k_m}\right\} \end{cases},$$

where every $\mathbf{x_{k_p}}$ is a variable not occurring in $\mathbf{P}$, and $\mathbf{x_i}$ is u_i for $u_i \notin T(\Phi)$.
3. Otherwise,

$$I_F(\neg\Phi) = \neg(I_F(\Phi)),$$

$$I_F(\Phi_1 \wedge \Phi_2) = (I_F(\Phi_1)) \wedge (I_F(\Phi_2)),$$

$$I_F(\forall \mathbf{x}\Phi) = (\forall \mathbf{x})(I_F(\Phi)).$$

Since $T(\Phi)$ eventually becomes empty, I_F is well-defined.

Let us consider an example of F-morphisms.

Example 5.4 We can define an F-morphism of $(N; \leq)$ to $(N; +)$, where N is the set of natural numbers, $\leq$ the linear ordering on N and $+$ addition. If we define Bas by $\text{Bas}(\leq) = (\exists \mathbf{z})(\mathbf{x} + \mathbf{z} = \mathbf{y})$ and I_O by the identity, then $\langle I_O, I_F \rangle$ with Bas is an F-morphism of $(N; \leq)$ to $(N; +)$.

We should notice that the function I_F is "automatically" defined according to the definition if I_O with Bas is already defined. Furthermore it is clear that if $(N; \leq) \vDash \phi$, then$(N; +) \vDash I_F(\phi)$. For example, let ϕ be a sentence $(\forall \mathbf{xyz})(\mathbf{x} \leq \mathbf{y} \wedge \mathbf{y} \leq \mathbf{z} \to \mathbf{x} \leq \mathbf{z})$.

Then

$$\begin{aligned} I_F(\phi) &= I_F((\forall \mathbf{xyz})(\mathbf{x} \le \mathbf{y} \wedge \mathbf{y} \le \mathbf{z} \rightarrow \mathbf{x} \le \mathbf{z})) \\ &= (\forall \mathbf{xyz})(I_F(\mathbf{x} \le \mathbf{y}) \wedge I_F(\mathbf{y} \le \mathbf{z}) \rightarrow I_F(\mathbf{x} \le \mathbf{z})) \\ &= (\forall \mathbf{xyz})((\exists \mathbf{z}_1)(\mathbf{x} + \mathbf{z}_1 = \mathbf{y}) \wedge (\exists \mathbf{z}_2)(\mathbf{x} + \mathbf{z}_2 = \mathbf{y}) \rightarrow (\exists \mathbf{z}_3)(\mathbf{x} + \mathbf{z}_3 = \mathbf{y})). \end{aligned}$$

So $(N; +) \models I_F(\phi)$.

Example 5.5 [Automaton] A Moore type automaton, $M_r = (A, B, C, \phi_r, \lambda_r)$ with A, B, C: finite sets, $\phi_r : C \times A \rightarrow C$ and $\lambda_r : C \rightarrow B$, is regarded as equivalent to a Mealy type automaton, $M_e = (A, B, C, \phi_e, \mu_e)$ with A, B, C: finite sets, $\phi_e : C \times A \rightarrow C$ and $\mu_e : C \times A \rightarrow B$. Let

$$\text{Moore} = \left\langle A \cup B \cup C;\ A,\ B,\ C, \hat{\phi}_r, \hat{\lambda}_r \right\rangle$$

and

$$\text{Mealy} = \left\langle A \cup B \cup C;\ A,\ B,\ C, \hat{\phi}_e, \hat{\mu}_e \right\rangle,$$

where A, B, C: unary relations, $\hat{\phi}_r$, $\hat{\lambda}_r$, $\hat{\phi}_e$, $\hat{\mu}_e$: arbitrary extensions of ϕ_r, λ_r, ϕ_e, μ_e respectively, and $\hat{\phi}_r = \hat{\phi}_e$, $\hat{\mu}_e(c, a) = \hat{\lambda}_r\left(\hat{\phi}_r(c, a)\right)$ for any $c, a \in A \cup B \cup C$.

Then we can define an F-morphism of Mealy to Moore as follows.

$$\left. I_O : \text{the identity} \right\},$$

$$\text{Bas}\left(\hat{\phi}_e\right) = \hat{\phi}_r(\mathbf{x}, \mathbf{y}, \mathbf{z}),$$

$$\text{Bas}\left(\hat{\mu}_e\right) = \left(\hat{\lambda}_r\left(\hat{\phi}_r(\mathbf{x}, \mathbf{y})\right) = \mathbf{z}\right),$$

$$\text{Bas}(\mathbf{A}) = \mathbf{A}, \quad \text{Bas}(\mathbf{B}) = \mathbf{B}, \quad \text{Bas}(\mathbf{C}) = \mathbf{C}.$$

These definitions obviously satisfy the conditions of basic morphisms. The following corollary shows that an F-morphism is an extension of a homomorphism.

Corollary 5.1 *Let $\mathfrak{M}_1$ and $\mathfrak{M}_2$ be system models of the same type, and h a homomorphism of $\mathfrak{M}_1$ to $\mathfrak{M}_2$. Then h is a basic morphism of $\mathfrak{M}_1$ to $\mathfrak{M}_2$, and $\langle h, I_F \rangle$ is an F-morphism of $\mathfrak{M}_1$ to $\mathfrak{M}_2$, where I_F is a function uniquely determined by h in Definition of F-morphism.*

From the definition of basic morphism we can see that an F-morphism preserves generators. Notice that it is necessary to give an interpretation I_F of the generators in defining an F-morphism, while in the case of a homomorphism I_F is trivially defined, and is not explicitly given in usual algebra.

Preservation of Σ: Σ_F-Morphism

In this section we will define an F-morphism preserving axioms Σ as a Σ_F-morphism. By a Σ_F-morphism, Σ is preserved in an image of an F-morphism. First we define the image of a basic morphism.

Let $\mathfrak{M}_1$ and $\mathfrak{M}_2$ be system models of possibly different types. Let $I = \langle I_O, I_F \rangle : \mathfrak{M}_1 \rightarrow \mathfrak{M}_2$ be an F-morphism. Then *the image of a basic morphism*, written $I(\mathfrak{M}_1)$, is defined by:

$$I(\mathfrak{M}_1) = \left\langle I_O(M_1); \left\{ R_i^2 \cap I_O(M_1)^{\lambda_2(i)} | i \in I_2 \right\}, \right.$$
$$\left\{ f_j^2 \parallel I_O(M_1)^{\mu_2(j)} | f_j^2 \left(a_1, \ldots, a_{\mu_2(j)} \right) \in I_O(M_1) \right.$$
$$\left. \times \text{ for any } a_1, \ldots, a_{\mu_2(j)} \in I_O(M_1), j \in J_2 \right\} \Big\rangle .$$

Notice that if there exists a j such that $f_j^2(a_1, \ldots, a_{\mu_2(j)}) \notin I_O M_1$ for $a_1, \ldots, a_{\mu_2(j)} \in I_O(M_1)$, then $I(\mathfrak{M}_1)$ is of a different type from $\mathfrak{M}_2$. An F-morphism is called an *onto F-morphism* if its basic morphism is onto.

Definition 5.13 [Σ_F-Morphism] Let $\mathfrak{M}_1$ and $\mathfrak{M}_2$ be system models and $I = \langle I_O, I_F \rangle$ an F-morphism of $\mathfrak{M}_1$ to $\mathfrak{M}_2$. Suppose $\mathfrak{M}_1 \models \Sigma$. Then I is called a Σ_F-*morphism* of $\mathfrak{M}_1$ to $\mathfrak{M}_2$ if

$$I(\mathfrak{M}_1) \models I_F(\Sigma),$$

where $I_F(\Sigma) = \{I_F(\Phi) | \Phi \in \Sigma\}$.

A Σ_F-morphism is a kind of extension of a Σ_F-homomorphism. The following corollary says that a Σ_F-morphism between system models of the same type accords with a Σ-homomorphism.

Corollary 5.2 *Let $\mathfrak{M}_1$ and $\mathfrak{M}_2$ be system models of the same type. Let $h : \mathfrak{M}_1 \rightarrow \mathfrak{M}_2$ be a homomorphism of $\mathfrak{M}_1$ to $\mathfrak{M}_2$. Suppose $\mathfrak{M}_1, \mathfrak{M}_2 \models \Sigma$. Then h is a Σ-homomorphism of $\mathfrak{M}_1$ to $\mathfrak{M}_2$ if and only if $\langle h, I_F \rangle$ is a Σ_F-morphism of $\mathfrak{M}_1$ to $\mathfrak{M}_2$.*

Preservation of Th ($\mathfrak{M}$): S_F-Morphism

In this section we define a morphism between system models of different types, which preserves Th ($\mathfrak{M}$) of a system model. Furthermore we will show the F-morphism theorem corresponding to the homomorphism theorem in the case of the same type. The F-morphism theorem gives a relationship between an F-morphism and an S_F-morphism. Unless mentioned explicitly, in the sequel let $\mathfrak{M}_1$, $\mathfrak{M}_2$ be system models and $I = \langle I_O, I_F \rangle$ an F-morphism of $\mathfrak{M}_1$ to $\mathfrak{M}_2$.

Definition 5.14 [S_F-Morphism] An F-morphism I is called an *S_F-morphism* of $\mathfrak{M}_1$ to $\mathfrak{M}_2$ if for any sentence Φ of $\mathfrak{L}(\mathfrak{M}_1)$, $\mathfrak{M}_1 \models \Phi$ if and only if $I(\mathfrak{M}_1) \models I_F(\Phi)$.

From the definition an S_F-morphism is a Σ_F-morphism. In general, we can check whether an F-morphism is an S_F-morphism by the way based on a structural induction on sentences. However in most cases, they are more than routine tasks.

For an onto F-morphism I, we define t*he quotient system model with respect to I* by:

$$\mathfrak{M}_1/I = \left\langle M_1/I_O : \left\{ R_i^1/I | i \in I_1 \right\}, \left\{ f_j^1/I | j \in J_1 \right\} \right\rangle,$$

where M_1/I_O is the partitioned set of M_1 by the equivalence relation $\equiv_{I_O}$ defined by:

$$a \equiv_{I_O} b \quad \text{if and only if} \quad I_O(a) = I_O(b) \quad \text{for any} \quad a, b \in \mathbf{M_1},$$

$$R_i^1/I\left([a_1], \ldots, [a_{\lambda_1(i)}]\right) \quad \text{if and only if Bas}\left(R_i^1\right)\left(I_O(a_1), \ldots, I_O\left(a_{\lambda_1(i)}\right)\right),$$

$$f_j^1/I\left([a_1], \ldots, [a_{\mu_1(j)}]\right) = \left[f_j^1\left(a_1, \ldots, a_{\mu_1(j)}\right)\right],$$

where $[a]$ represents an equivalence class in M_1/I_O by $\equiv_{I_O}$.

The above definition of f_j^1/I is well-defined. Indeed, suppose $I_O(a_1) = I_O(b_1), \ldots, I_O\left(a_{\mu_1(j)}\right) = I_O\left(b_{\mu_1(j)}\right)$ and $f_j^1\left(a_1, \ldots, a_{\mu_1(j)}\right) = c_1$, $f_j^1\left(b_1, \ldots, b_{\mu_1(j)}\right) = c_2$ hold in $\mathfrak{M}_1$. Then since I is an F-morphism, $\text{Bas}\left(\mathbf{f_j^1}\right)\left(I_O(a_1), \ldots, I_O\left(a_{\mu_1(j)}\right), I_O(c_1)\right)$ and $\text{Bas}\left(\mathbf{f_j^1}\right)\left(I_O(b_1), \ldots, I_O\left(b_{\mu_1(j)}\right), I_O(c_1)\right)$ hold in $\mathfrak{M}_2$. Since $\text{Bas}\left(\mathbf{f_j^1}\right)$ is a function from the definition of basic morphism, we have $I_O(c_1) = I_O(c_2)$.

The following is one of the main theorems about F-morphisms.

Theorem 5.2 [F-Morphism Theorem] *Let $I : \mathfrak{M}_1 \to \mathfrak{M}_2$ be an onto F-morphism. Then*

$$I_O^{\#} : \mathfrak{M}_1/I \to \mathfrak{M}_2$$

is a one-to-one basic morphism, furthermore

$$I^{\#} = \left\langle I_O^{\#}, I_F^{\#} \right\rangle$$

is an S_F-morphism, where $I_O^{\#}$ is defined by

$$I_O^{\#}([a]) = I_O(a) \quad \text{for} \quad [a] \in M_1/I_O$$

and the basic interpretations are

$$\text{Bas}^{\#}\left(\mathbf{R}_\mathbf{i}^\mathbf{1}/I\right) = \text{Bas}\left(\mathbf{R}_\mathbf{i}^\mathbf{1}\right) \quad \text{for } i \in I_1,$$

$$\text{Bas}^{\#}\left(\mathbf{f}_\mathbf{j}^\mathbf{1}/I\right) = \text{Bas}\left(\mathbf{f}_\mathbf{j}^\mathbf{1}\right) \quad \text{for } i \in J_1.$$

$I^{\#}$ is called *the induced F-morphism* of *I*.

F-morphism enhances the concept of "isomorphism" between system models to those of different types, which means that the two models described in "different languages" are isomorphic in terms of F-morphism.

5.3.4.3 Application of F-Morphisms

A typical way to apply the F-morphism concept to concrete system models would be to construct an F-morphism between them. In this section, as an application of F-morphisms, we construct an F-morphism of a given finite automaton structure to a Petri net structure, and show the equivalence between a finite automaton and a Petri net. The equivalence means here that we can show that there is a Petri net that preserves all the properties of a given finite automaton. It is well known that Petri nets can represent finite automata (Peterson, 1981). The emphasis in this section, however, is on that the use of the F-morphism concept in considering the equivalence between a finite automaton and a Petri net reveals that each property of the finite automaton precisely (in a formal way) corresponds to some property of the Petri net, and the first-order sentences satisfied in the finite automaton are all preserved in the Petri net as corresponding sentences transformed by an F-morphism. Thus the F-morphism concept provides a formal meaning of "equivalence" between system models of different types, while the judgment of the equivalence "without F-morphisms" would depend fully on the intuition of a modeler constructing the correspondence between them.

Equivalence Between a Finite Automaton and a Petri Net

In this section we construct an F-morphism of a given finite automaton structure to a Petri net structure, and show that all the properties holding in the finite automaton also hold in the Petri net.

Definition 5.15 A finite automaton structure FA is the following system model.

$$\text{FA} = \langle A \cup B \cup C;\, A, B, C, \phi, \rho \rangle ,$$

where

A, B, C: unary relations
ρ: binary functions such that

$\phi(a, b) \in C$, if $a \in C$ and $b \in A$
$\phi(a, b) = a$, otherwise;
and $\rho(a, b) \in B$, if $a \in C$ and $b \in A$
$\rho(a, b) = a$, otherwise.

The conditions on $a \notin C$ or $b \notin A$ for ϕ and ρ are imposed only to make the functions ϕ and ρ total, since the first-order language we use does not allow partial functions. However, since we will restrict the sentences to the extent as defined later, when we describe the properties of system models, we can regard ϕ and ρ intrinsically as $\phi : C \times A \to C$ and $\rho : C \times A \to B$.

Definition 5.16 [Petri Net Structure] A Petri net structure PN is the following system model.

$$\text{PN} = \left\langle P \cup T \cup N;\, P, T, I, O, \hat{N} \right\rangle ,$$

where

P, T: unary relations
N: the set of natural numbers
$\hat{N}$: the set of constants corresponding to N
$I, O \subset P \times T \times N$

P denotes the set of places and T the set of transitions. $I(p, t, n)$ means that there are n arcs from the place p to the transition t. $O(p, t, n)$ means that there are n arcs from the transition t to the place p.

There are some ways to construct PN that is considered to have an equivalent structure to FA (Peterson, 1981). Here following Peterson with some modification, we define PN considered as equivalent to FA. Then our aim is to construct an F-morphism between FA and PN, and to show that the constructed PN preserves all the properties satisfied in FA.

Definition 5.17 Given a finite automaton structure FA. We define the corresponding Petri net structure PN as follows.

$$\text{PN} = \left\langle P \cup T \cup N;\, P, T, I, O, \hat{N} \right\rangle ,$$

where

$P = C \cup A \cup B;$
$T = \{t_i | i \in (C \times A) \cup A \cup B\};$
$I = I_1 \cup I_0,$

where

$I_1 = \{(p, t_i, 1) | i = (c, a) \in C \times A \wedge (p = c \vee p = a)\} \cup \{(p, t_i, 1) | i = p \in B\},$
$I_0 = \{(p, t_i, 0) | (p, t_i, 1) \notin I_1, p \in P, t_i \in T\},$
$O = O_1 \cup O_0,$

where

$O_1 = \{(p, t_i, 1) | i = (c, a) \in C \times A \wedge (p = \phi(c, a) \vee \rho(c, a))\} \cup \{(p, t_i, 1) | i = p \in A\},$
$O_0 = \{(p, t_i, 0) | (p, t_i, 1) \notin O_1, p \in P, t_i \in T\}.$

Definition 5.18 Let FA and PN be as in the above definitions, respectively. An F-morphism $I = \langle I_O, I_F \rangle$: FA $\rightarrow$ PN is defined as follows.

I_O: the inclusion map;
$I_F(\mathbf{A(x)}) = (\mathbf{P(x)} \wedge (\exists \mathbf{t} \in \mathbf{T})((\forall \mathbf{p} \in \mathbf{P})(\mathbf{I(p, t, 0)} \wedge \mathbf{O(x, t, 1)})));$
$I_F(\mathbf{B(x)}) = (\mathbf{P(x)} \wedge (\exists \mathbf{t} \in \mathbf{T})((\forall \mathbf{p} \in \mathbf{P})(\mathbf{O(p, t, 0)} \wedge \mathbf{I(x, t, 1)})));$
$I_F(\mathbf{C(x)}) = (\mathbf{P(x)} \wedge \neg I_F(\mathbf{A(x)}) \wedge \neg I_F(\mathbf{B(x)}));$
$I_F(\phi(\mathbf{x}, \mathbf{y}) = \mathbf{z}) = \Big((I_F(\mathbf{C(x)}) \wedge I_F(\mathbf{A(y)}) \rightarrow (\exists \mathbf{t} \in \mathbf{T})(\mathbf{I(x, t, 1)} \wedge \mathbf{I(y, t, 1)}$
$\wedge \mathbf{O(z, t, 1)} \wedge I_F(\mathbf{C(z)})))) \wedge (\neg I_F(\mathbf{C(x)}) \vee \neg I_F(\mathbf{A(y)}) \rightarrow \mathbf{z} = \mathbf{x}) \Big);$
$I_F(\rho(\mathbf{x}, \mathbf{y}) = \mathbf{z}) = \Big((I_F(\mathbf{C(x)}) \wedge I_F(\mathbf{A(y)}) \rightarrow (\exists \mathbf{t} \in \mathbf{T})(\mathbf{I(x, t, 1)} \wedge \mathbf{I(y, t, 1)}$
$\wedge \mathbf{O(z, t, 1)} \wedge I_F(\mathbf{B(z)})))) \wedge (\neg I_F(\mathbf{C(x)}) \vee \neg I_F(\mathbf{A(y)}) \rightarrow \mathbf{z} = \mathbf{x}) \Big).$

This definition clearly satisfies the condition required for F-morphisms. Also we can see, as the following lemmas show, that the image of the above F-morphism preserves the structure of FA.

The following theorem shows a typical type of equivalence between PN and FA.

Theorem 5.3 [Equivalence of the Structures of Finite Automaton and Petri Net] Let $I = \langle I_O, I_F \rangle$ be the F-morphism defined as in the above Definition. Then for any many-sorted sentence Φ of $\mathfrak{L}$(FA), FA $\models \Phi$ if and only if PN $\models I_F(\Phi)$.

This theorem implies that the structure of FA is embedded in PN constructed in Definition, and all the properties of FA are preserved there. We should notice that the dynamic behavior of PN by the transition of marking is implied by the relation O of PN, which can also represent the firing of the transitions.

5.4 Structure and Adaptation

5.4.1 *Implications of Gödel's Incompleteness Theorem for Structural Change*

The definition of the structure of a system model implies that there are two ways to represent a change of the structure: the changes of the symbols $\mathfrak{L}$ and of the axioms Σ. Here we deal with the change of the axioms. Then we can naturally say that a system has changed its structure from Σ to $\Sigma^{'}$, if $\Sigma^{'} = \Sigma + \{\psi\}$, where ψ nor $\neg\psi$cannot be derived from Σ, i.e.,$\Sigma \vdash \psi$ nor $\Sigma \vdash \neg\psi$, under some deduction system.

This new property ψ cannot be recognized in the old structure Σ. Classical deduction systems such as first-order logic cannot deal with this situation effectively. Also in ordinal systems theory based on set theory without logical language and deduction system we hardly develop comprehensive consideration on this matter. We need to construct a meta-framework in which the structure of a system ($\mathfrak{L}$; Σ) can be referred an "object" and to extend the concept of system model by adding extra domain to it.

Gödel proved incompleteness theorem that there is a sentence that cannot be inferred from the logical system including the primitive recursive arithmetic (PRA) such as Peano's axioms. From the Gödel's incompleteness theorem, we see that there is a formula such that PRA $\vdash \phi \leftrightarrow \neg \Box \phi$. The symbol $\Box$ expresses a modal operator and the formula $\Box\phi$ is interpreted to be "ϕ is provable" (Symoryński, 1985). Then $\neg \Box \phi$ means that ϕ is not provable. Furthermore $\phi \leftrightarrow \neg \Box \phi$ means that "I am not provable." This formula shows a self-referential sentence. Hence if the structure of a system includes PRA, the system has a property ϕ such that ϕ cannot be implied from the structure, and nor identified as behavior of the system.

As stated in the previous paragraph, if the structure of a system "moves" to a new structure that includes the old structure and the above self-referential sentence ϕ that is never proved from the old one, the system actually "change" its structure: the move from the structure Σ to $\Sigma^{'} = \Sigma + \{\phi\}$ represents a structural change of the system.

5.4.2 *Adaptation in Social System: Agent-Based Organizational Cybernetics*

There are two kinds of adaptive behavior of a system: first-order adaptive behavior and second-order adaptive behavior. The first-order adaptive behavior seeks to make the system stable by regulating the gap between the goal of the system and the systems output observed. It is realized as a negative feedback mechanism, which can be formulated in the structure of the system, i.e., in LAST. The second-order adaptive behavior requires to change the field of systems behavior, which can be represented as the concept of positive feedback or second-order cybernetics. The second-order adaptive behavior cannot be described as any properties derived from the structure of the system. As seen in the above, the self-referential sentence plays a key role in the adaptation by the structural change. We need to introduce some new conceptual devices to describe models of the second-order adaptive behavior: agent, internal model, and organizational learning.

This section focuses on adaptation in social systems and provides the theoretical framework for describing it. Here a comprehensively hybrid model is introduced, which combines conventional organizational cybernetic framework and computational organization theoretic approach, especially agent-based computational learning model. The framework of organizational cybernetics includes no agent concept innately, but originally aims to contribute to the diagnosis of organizational failure based on Ashby's law of variety (Espejo et al., 1996). On the other hand, computational organization theoretic approach contains agent-based task resolution processes in detail operational manner, but describes only a "flat" organization that has no hierarchical relationship between subsystems. The hybrid model presented here is comprehensive in understanding organizational learning in the sense that the learning process includes essentially the following steps: each agent resolves tasks in every functional layer in an organization; the results of the resolutions of tasks are unified to be organizational output performance; the organizational performance should be evaluated from environment; each agent change its internal model based on the evaluation results.

We call our newly developed approach Agent-based Organizational Cybernetics (AOC) (Takahashi, 2006). An organization considered in AOC is formulated to have four functional layers defined in organizational cybernetics: process, coordination, adaptation, and self-organization. The organizational cybernetic model has originally no concept of an agent. AOC introduces the concepts of agent and communication process among agents into each functional layer of an organizational cybernetic model. An agent is characterized by individual situatedness and internal model principle. The agent is defined as an autonomous decision-maker who constructs individually its internal model to describe its recognition of the situation surrounding it.

The basic features of AOC can be listed below.

1. Interaction between environment and decision-makers (from organizational cybernetic viewpoint).
2. An autonomous decision-maker makes a decision according to his decision principle.
3. An organization is structured in a multi-layer hierarchical form with some functional subsystems.
4. In each layer of the hierarchy, some agent groups are involved and interact one another.
5. Every agent has its own internal model that describes the situation surrounding the agent (called individual situatedness).

6. Every agent can learn its internal model and the organization can learn by sharing agents' internal models. The process of learning represents single- and double-loop learnings in organizational learning.

AOC allows us to deal with organizational problems such as organizational learning in essentially operational manners. Results analyzed using AOC could suggest how we should effectively and operationally manage complex problems on the organization concerned. The principal target of AOC is to provide design criteria of prescription, especially which has not yet been validated in actual situations, on how an organization of concern should make a decision and take an action to adapt itself to a dynamically changing environment. AOC can also provide an effective way to evolve new design of functions working in an organization by re-combination of subsystems.

5.4.3 *Components of Organizational Learning*

The concept of organizational learning we use for our framework has similar aspects to the Argyris' concept that individual learning processes are innately connected to organizational learning process. Our framework explicitly distinguishes individual levels of learning and organizational ones, and also does the levels of single-loop learning and double-loop one. We can see the explicit distinctions of the four types of learning loops.

The distinction by Argyris of single-loop learning and double-loop one is originated, as Argyris stated (Argyris & Schön, 1996), from similar notions in cybernetics developed by Ashby (1960). Based on organizational cybernetic approach, Espejo has provided a basis for the way how to apply the double-loop learning notion and process to actual organizations.

In AOC the concept of organizational learning, especially the learning-loop processes are realized in operational ways that each agent evaluates and revises its internal model. By actually implementing mechanisms of organizational learning in agents, the micro-macro problems can be explored effectively so as to tackle complex organizational systems.

The essential elements of organizational learning in AOC are the four learning loops: individual−/organizational- and single−/double-loop learning. AOC implements the learning processes as evaluating, revising, and sharing processes of internal models possessed by agents.

1. Individual single-loop learning.

   An agent builds its internal model to describe the environmental structure and the problem situation recognized, which includes some decision variables and decision criteria. The agent uses its internal model to optimize the decision variables. This learning does not enhance any ability to make organizational decisions.
2. Organizational single-loop learning.

To achieve the given organizational goal, subgoals are specified to agents in inferior subsystems The values of the individual decision variables, which must be the results of the individual single-loop learning, are unified by the organization. The organization makes a decision based on the unified results.

3. Individual double-loop learning.

   Each agent evaluates its internal model, based on the results of the decision performed just before. Then the internal model is revised.

   This process of revising internal models by agents can be implemented effectively by using genetic algorithm (GA). The evaluation is defined by a fitness function that indicates what kind of information is available and how it should be utilized for the evaluation.

   After the evaluations of the internal models, applying GA operators such as selection, crossover, and mutation, the internal models are revised for the subsequent decisions.

4. Organizational double-loop learning.

   As the result of "good" individual double-loop learning, the agents share in the organization their good internal models that provide them with better decision capability and allow them to keep the organization viable.

5.5 Basis of Agent-Based Organizational Cybernetics

5.5.1 Hierarchical Organization Model in AOC

Combining the hierarchical model of organizational cybernetics and the agent model in computational organization theory, AOC consists of two basic models: hierarchical organization model and situated agent model.

A hierarchical organization model is a multi-layer system that has basically adaptive, coordination, and operational levels (Fig. 5.1).

In AOC the function of each level is realized by a group of agents. Every agent belongs to one of the subsystems of the hierarchy. Each subsystem seeks a possibly different goal from other subsystems. Hence an agent is conducted based on the goal of the subsystem of which the agent is a member.

The adaptive level is composed of intelligence and institutional functions. In this level, based on environmental information observed by the intelligence function, the organization creates policy or strategy that could achieve a given organizational goal. If the organizational goal is recreated, the organization would go to a self-organization phase.

The coordination level has a function that determines coordination variables to control inferior subsystems in a decentralized manner. Coordination principles, which define how to coordinate the inferior subsystems, are essential to achieve a coordination goal.

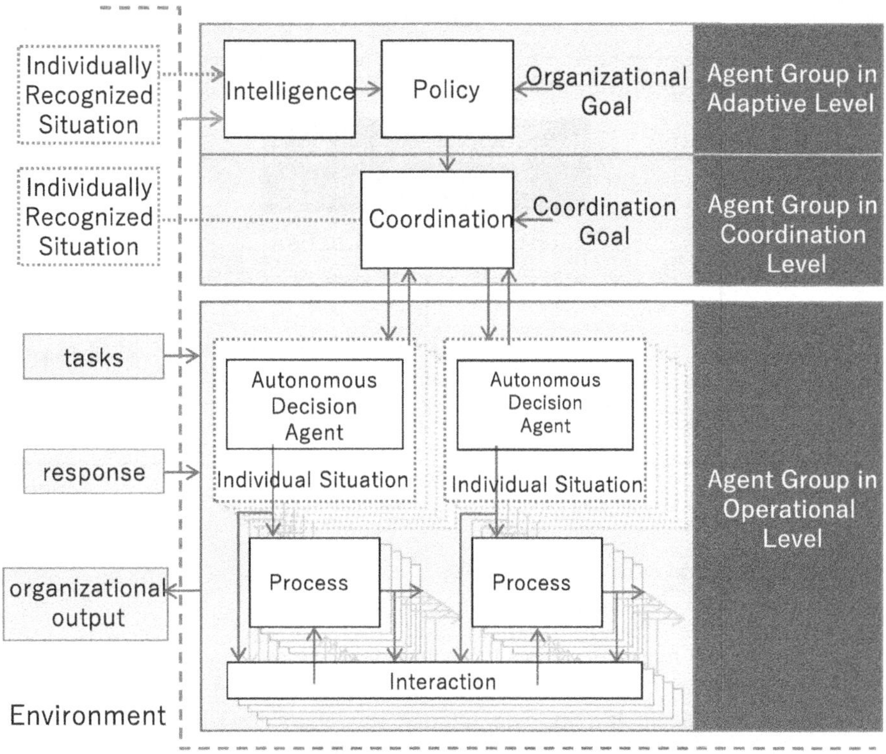

Fig. 5.1 Basic hierarchical model in AOC

In the operational level, agents determine decision variables autonomously, each of whom aims to optimize the process assigned to him. The processes interact with one another. The optimization process is given as a task resolution one, the result of which is reported to the superior subsystem, i.e., coordination level.

Computational Organization Theory has focused so far on models of the operational level. AOC formulates the operational level as a layer of the hierarchical subsystems of the overall organization model.

5.5.2 Situated Agent Model in AOC as Autonomous Decision-Maker

An agent concept in AOC as an autonomous decision-maker has basically the following features (Fig. 5.2).

1. An agent recognizes a process as a target of its decision-making, and builds its model internally, which is called an internal model.

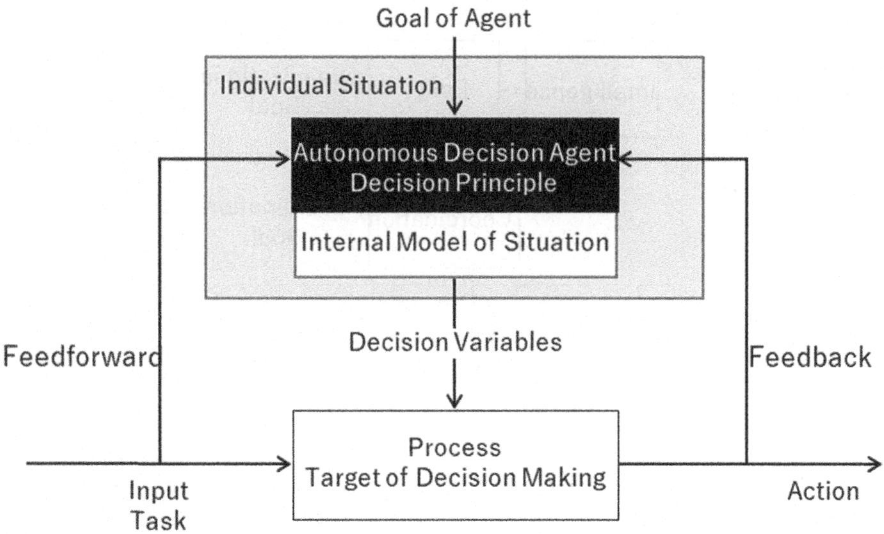

Fig. 5.2 Situated agent model as decision-maker

The internal model describes the behavior of the process and external inputs from environment, which can include agent's recognition of the surrounding situation. An agent has its own internal model to describe the surrounding situation. Every agent is considered to be involved in its situation that is individually perceived by that agent. We call it individual situatedness.

2. An agent applies a decision principle to a problem concerning the process so that the agent evaluates options or alternatives to solve the problem.

The decision principle represents a criterion for preference ordering of the alternatives. It can be formally defined as a function from the class of problems to be solved into the ordering structures of preferences.

An eminent feature of AOC is to deal with, in an operational manner, micro-macro link problems such as a problem of the relationship between individual learning process of each agent and organizational one. An agent is typically a member of one of the autonomous decision groups defined in the multi-layers of an organization. The overall environment can be recognized as interpreted information from shared internal models of individually perceived situations.

5.5.3 *Typical Internal Models*

We here consider typical internal models in each hierarchical level.

A typical internal model that an agent in adaptive level has its recognition of the environmental structure, especially the recognition how the environment makes responses to an agent and the organization in the form of cost-profit function.

Another typical internal model in this level is the decision principle that an agent uses to make its decision. An internal model used in coordination level is the recognition of the process in which assigned tasks should be actually resolved.

In the operational level, a typical internal model can be how an agent recognizes tasks to be resolved as well as the task resolution process itself.

The point in considering learning problems in an organization is how each agent should evaluate its internal model, based on which the agent revises its internal model, i.e., the recognition of its individual situation and shares it among agents.

5.6 Conclusion: Beyond LAST

Logical Approach to Systems Theory (LAST) provides a way to investigate the similarity and structure of system models in type-free representation. Hence LAST clarifies the similarity of systems models in different representations as "isomorphic relation" with F-morphism. LAST explicitly gives the concept of the structure of a system model as a pair of the description language and the axioms characterizing the system model. Then LAST clarifies the distinction of what type of properties of systems can be described in the theory and what type of them cannot. Adaptation, which has been a central concern of systems theory, is a typical type of properties logical approaches hardly describe. One possibility of describing adaptation in logical frameworks would be to exploit modal logic, which could express the self-referential concept. Then the concepts of agent, internal model, and organizational learning would provide a breakthrough way to describe the adaptation of systems.

Agent-based organizational cybernetics involves as its significant part an adaptive mechanism in which agents revise intrinsically their own internal models and then share them with the other agents of the system through the organizational learning process the system constructs. The adaptive mechanism has been developed using many types of evolutionary computation algorithms such as genetic algorithm and so on. Tons of research results concerning them have been accumulated. Our future target would construct a way to formulate such adaptive processes as clearly as possible based on these outcomes, which would give a landscape beyond LAST.

References

Argyris, C., & Schön, D. A. (1996). *Organizational learning II*. Addison-Wesley.
Ashby, W. R. (1960). *Design for a brain* (2nd ed.). Wiley.
Chang, C. C., & Keisler, H. J. (1973). *Model theory*. North-Holland.

Espejo, R., Schuhmann, W., Schwaninger, M., & Bilello, U. (1996). Organizational transformation and learning. In *A cybernetic approach to management*. Wiley.
Grätzer, G. (1979). *Universal algebra*. Springer.
Mesarovic, M. D., & Takahara, Y. (1975). *General systems theory: Mathematical foundation*. Academic Press.
Mesarovic, M. D., & Takahara, Y. (1989). *Abstract systems theory*. Springer.
Peterson, J. L. (1981). *Petri net theory and the modeling of systems*. Prentice-Hall.
Symoryński, C. (1985). *Self-reference and modal logic*. Springer.
Takahashi, S. (1995). Self-referential systems representation with modality in structure change. *Advances in Systems Science and Applications*, Special Issue, The International Institute for General Systems Studies, pp 19–24.
Takahashi, S. (2006). Agent-based organizational cybernetic approach to organizational learning. In *SICE-ICASE International Joint Conference 2006(SICE-ICCAS 2006)*.
Takahashi, S., & Takahara, Y. (1995). *Logical approach to systems theory*. Springer.
Zeigler, B. P. (1990). Object-oriented simulation with hierarchical. In *Modular models: Intelligent agents and endomorphic systems*. Academic Press.

Chapter 6
Logical and Algebraic Structure of "Calculus of Indication": The Significance and Circumstance

Takuhei Shimogawa

Abstract This paper presents the statements with respect to G.Spencer Brown's theme, namely, Primary Algebra and Brownian Algebra, under standard mathematics. G.Spencer Brown's "Calculus of Indication" and "Primary Algebra" once attracted many researchers of various kinds of areas including mathematics, economics, sociology, etc. It looks finally that, despite the idea itself was somewhat new, the essence was said to be not new, rather similar to classical propositional calculus (without any modality) or Boolean algebra.

This paper gives formal proof to state the properties that the theme inevitably contains. The author presents a couple of "stories" to clarify the ambiguity and investigates real properties of Primary Algebra (Spencer Brown, Laws of form. George Allen and Unwin, 1969) and Brownian Algebra (Varela, Principle of biological autonomy. Elsevier North Holland, 1979). Some of the considerations concerning "interpretation" and "implication" are also given.

Keywords Calculus of indication · Primary arithmetic · Primary algebra · Brownian algebra · Inference system · Algebraic system · Universal algebra

6.1 Overview of the Subjects and Discussions

This paper takes the form of a re-enactment of an argument that has been criticized in the past and whose shortcomings have already been revealed, and a reconsideration of its significance.

Systems theory has evoked diverse debates in various contexts in the past. Among them, there are some systems-theoretic arguments in sociology and biology, which have been left kind of unconsidered and unclear in their implications.

T. Shimogawa (✉)
Faculty of Economics, Musashi University, Tokyo, Japan
e-mail: smgw@cc.musashi.ac.jp

K. Kijima et al. (eds.), *Systems Research I*, Translational Systems Sciences 26,
https://doi.org/10.1007/978-981-16-9937-5_6

It also seemed the literatures had the tendency to employ "mathematical" standpoint, which seemed to have been influenced by the structuralism of mathematics. Among the influenced contexts, the Calculus of Indication and Primary Algebra (Spencer Brown, 1969) seemed to have had some reputation and specialty. Therefore I made these "mathematical notions" targets of analysis.

The general pattern for treating or examining an ambiguous subject in neutral and logical way, providing some coherent methods to comprehend the subject (preferably more than one method) would be depicted as follows:

A certain number of results and discussions are presented, some of which may contribute to the resolution for our target, roughly based on this diagram. This paper could provide some examples of (a)–(g) for the purpose. To be concrete, the simple classical logic and abstract algebraic theory are used to formulate the notions.

First a "logic-like" reformulation of the subject(Primary Algebra) is introduced based on Shimogawa and Takara (1994). Then the second story is given, in which formal analysis of Brownian Algebra is demonstrated, with the comparison of Primary Algebra. Brownian Algebra was developed by F. J. Varela (1979).

As for the second story, several facts about the relation between algebraic formalization of Primary Algebra and Boolean algebra are clarified. Also, direct relations between algebraically formalized Brownian Algebra and distributive lattice are shown. Several results of noteworthy for them are presented. On demonstrating the second story, the notion of ordinary universal algebraic theory is used. The systems are formalized into algebras (Graetzer, 1979) in simple and straightforward way. The notions "weak algebra" and "implementation" are introduced. Finally they are used in order to establish the statements that contribute to the author's motivation mentioned in Fig. 6.1.

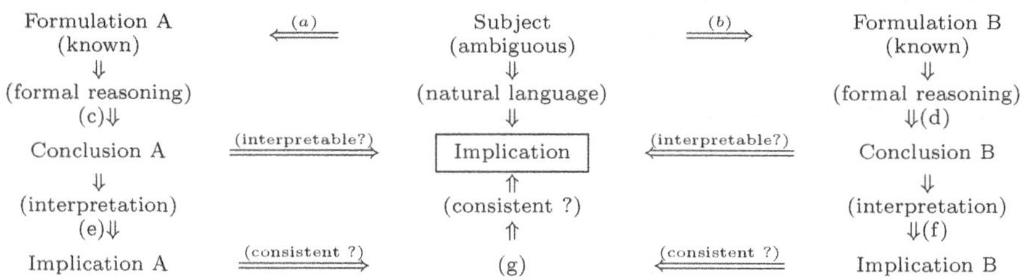

Fig. 6.1 General view of systems research

6.2 Reviewing G.Spencer Brown's Theme

6.2.1 Brief Introduction of G.Spencer Brown's Framework

In the introduction of "Primary Arithmetic" of the G.Spencer Brown's book (Spencer Brown, 1969), the design and idea were depicted in ambiguous and philosophical manner.

According to Spencer Brown (1969) (and Varela, 1979), an *indicational space* is given, in which the act of "distinction" occurs. "Primary Arithmetic" is defined to be a calculation that is done on the indicational space and has a pair of the rule of calculation called *arithmetic initials* described as follows:

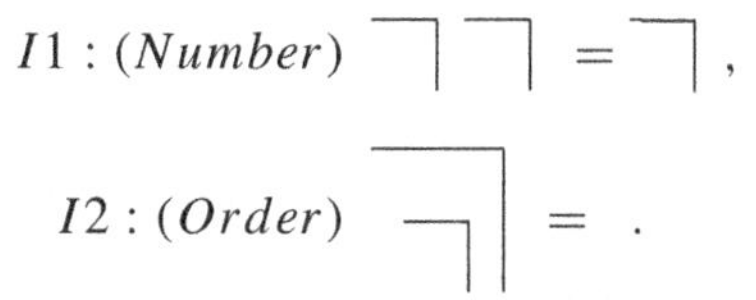

It should be noted that the equality symbol "=" occurring in I1 and I2 is not one that we used in usual mathematics. Instead it is "command" telling "do the calculation from the left side to the right" or "do it conversely." An object of calculation (it is a collection of enclosures used in I1 and I2) is called *the expression*. Putting an "enclosure" on the "space" is called *the cross*. Following is an example of the expression:[1]

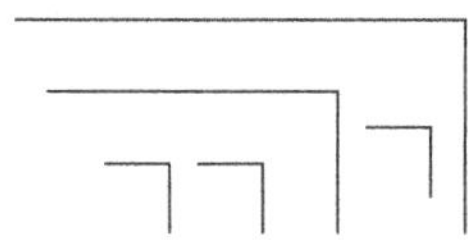

Using I1 and I2 as the rules, the calculation is executed as follows:

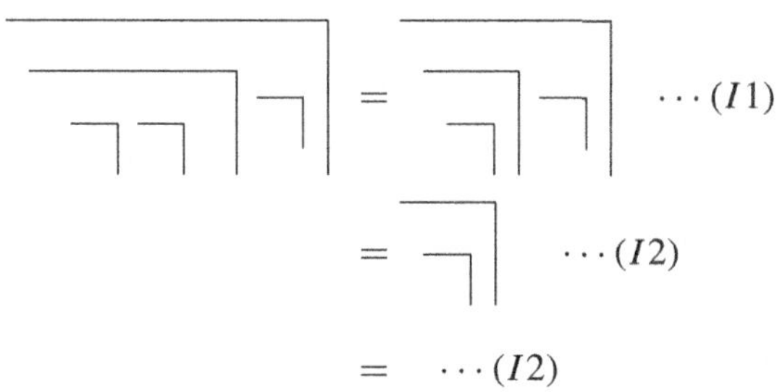

[1] As readers have noticed in this point, we are already puzzled: "What is the meaning of an expression ? Distinctions ? Do we distinguish things like that ?". Please refer to Spencer Brown (1969) for epistemological discussion. Setting them aside, let us proceed.

Briefly speaking, if we find $\overline{\ \ }|\,\overline{\ \ }|$ or $\overline{\overline{\ \ }|}\,|$ in a given expression, then we can apply I1 or I2 to reduce the crosses. And conversely we are allowed to add $\overline{\overline{\ \ }|}\,|$ everywhere (of course with no crossing) in the given expression or also allowed to add $\overline{\ \ }|$ to the side $\overline{\ \ }|$ if it occurs. If the "calculation" of simplification is not able to continue anymore, the resulted expression is called *simple expression*, and calculations obtaining simple expression or the simple expression obtained is called a *simplification*. It is pretty apparent to see that simple expressions are:

$\overline{\ \ }|$ or , and nothing else is simple expression.

(As was already seen, the blank: left element of "=" in I2, is somehow awkward to deal. We leave it not redefined for later section in which the definition of denotation of the blank will be introduced.)

FACT 2.1 (Spencer Brown, 1969) *The simplification of an expression is unique.* ■

Now it is a good timing to see the "hidden" definition or semantics of "=," as follows (of course based on the facts we have already seen):

Definition 2.2 For a couple of expressions: p and q,

$p = q$ iff p and q are simplified to the same simple expression.

As the simplification is always unique, we easily verify the following fact:

FACT 2.3 (Spencer Brown, 1969) *For any expressions p, q, and r, the followings hold:*

$$\overline{\overline{p}|\ \ p}\,| = \ ,$$

$$\overline{\overline{pr}|\ \overline{qr}|}\,| = \overline{\overline{p}|\ \overline{q}|}\,|\ \ r\ .$$

Two equations in Fact 2.3 are employed to construct "Primary Algebra." They are called *initials*, which word is unfamiliar in common mathematics. Let us follow the discussion done in Spencer Brown (1969). Later in this paper we will see the precise reconstruction (Shimogawa and Takara, 1994).

Definition 2.4 (Initials of Primary Algebra)

$$J1(Position): \quad \overline{\overline{p\,}\big|\;\; p\;}\Big| \;=\; ,$$

$$J2(Transposition): \quad \overline{\overline{pr\,}\big|\;\; \overline{qr\,}\big|\;}\Big| \;=\; \overline{\overline{p\,}\big|\;\; \overline{q\,}\big|\;}\Big|\;\; r\ .$$

It should be emphasized again that "Primary Algebra" is not an algebraic system in common sense. However en route we might see it as "reform-able" to a certain kind of algebraic system in which a couple of operations and its axioms are given. The equality "=" in "Primary Algebra" also seems to have quite a common meaning in usual algebraic systems, meaning that J1 and J2 indicate substitutability of the objects, for the left of the right, and vice versa.

To present an example of "Theorem" in "Primary Algebra" could be helpful for readers to comprehend G.Spencer Brown's intention in Spencer Brown (1969): The following "demonstration" (obviously the meaning is the derivation of an identity in algebraic systems) was given.

Conclusion of Reflection

$$\overline{\overline{p\,}\big|\;}\Big| \;=\; p\ .$$

Proof (In original book Spencer Brown (1969) the process is called "demonstration.")

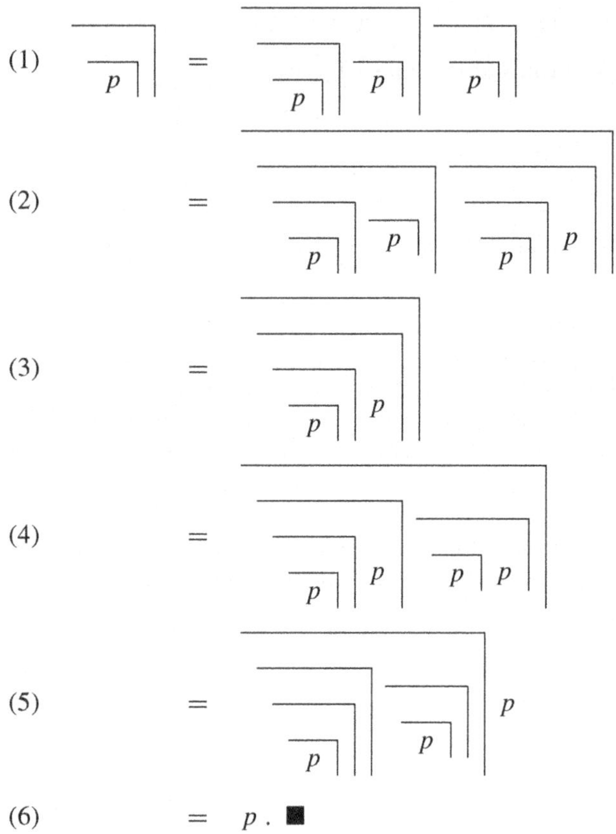

Remember that the term of the type 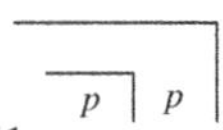 can be located everywhere (on the text line) the "blank" exists, due to J1.

6.2.2 *Direction*

The idea introduced in Spencer Brown (1969), which we have seen its introductory part above, was followed by a number of works in the system-theoretical context. Some of them we are not able to examine due to the fact that they are too philosophical and semantical. As for G.Spence Brown's theme itself and its direct application (Varela, 1979, for instance), the author thinks the total analysis under the formalism is possible.

The point is as follows: It is quite right that we feel not that hard to accept this "derivation." How does it occur, even though it is not mathematical ? As is well-known, the claim that it was based on the "calculation rules" that do not exist in any paradigm of mathematical context has been scathingly criticized. The criticism is apparently proper (as I present in this article). However, as far as we have noticed, accurate and detailed inspection for the whole idea derived from the original book has not been done yet.

For "Primary Algebra" we can easily prove all the "axioms" of it inside Boolean algebra if it is adequately formalized. As for Varela (1979), on the other hand, almost nothing has been done yet.[2]

As the author indicates later on, the "morphism" τ (we will see the definition in the following section) would be one of the subjects for the purpose of this article.

6.3 Story 1: Reconstruction of Primary Algebra As a Logical System

In this section, a demonstration of formalizing the subject into a *logical system* is presented. The redefinition of the Primary Algebra, called PA as a logical system is given. The story presented here relies entirely on Shimogawa and Takara (1994).

It is the first demonstration in this paper of transformation from "ambiguous/philosophical framework" to a formal system.

6.3.1 Definitions

Given a formal language L that consists of a set of characters and a set of production rules, a formal inference system S is defined to have a comprehensive method to decide the subset of well-formed formulas called the theory of S. In fact, the Primary Algebra can be treated as a kind of formal inference system. The following is a reformation of the PA originated in Schwartz (1979).

Definition 3.1 (Formal System PA)

- Primitive symbols of language for system PA (denoted by $L(PA)$):
  - variables: $\{e_i\}_{i \in J}$, $J \cong \omega$
  - blank symbol: $\{\varepsilon\}$
  - square brackets: $\{[,]\}$

[2] System theorists and researchers in other domains had lost their interest in the topic so quickly. The author thinks it is because crucial criticism against the so-called new-age science arouse during the 1990s.

- parenthesis: {(,)}
- binary predicate symbol: { == }

- Terms of $L(PA)$:

1. e_j $(j \in J)$ is a term.
2. ε is a term.
3. If E is a term then so is $[E]$.
4. If E_1 and E_2 are terms then so is $(E_1 E_2)$.
5. No exceptions.

The set of terms is denoted by Exp, or an element of Exp will be called "expression" following the original book (Spencer Brown, 1969).

- Formulas of $L(PA)$: If $E, F \in Exp$, then $E == F$ is a formula. We sometimes call a formula *an equation of expression*.
- Axioms: For any $E_1, E_2, E_3 \in Exp$,

A1: $[([E_1]E_1)] == \varepsilon$.
A2: $[([E_1E_3][E_2E_3])] == ([([E_1][E_2])]E_3)$.
A3: $(E_1\varepsilon) == E_1$.
A4: $(E_1E_2) == (E_2E_1)$.
A5: $((E_1E_2)E_3) == (E_1(E_2E_3))$.

- Subterm of Exp and its depth: For the technical reason, we introduce the notions of subterm and depth for an expression. The set of subterms of an arbitrary expression $G \in Exp$, $Sub(G) \subset Exp \times \mathbb{N}$ is defined as follows:

1. $\langle G, 1\rangle \in Sub(G)$.
2. If $\langle G', i\rangle \in Sub(G)$ and $G' = [F]$, then $\langle F, i+1\rangle \in Sub(G)$.
3. If $\langle G', i\rangle \in Sub(G)$ and $G' = G_1G_2$ for some $G_1, G_2 \in Exp$, then $\langle G_1, i\rangle \in Sub(G)$ and $\langle G_2, i\rangle \in Sub(G)$.
4. Nothing else is in $Sub(G)$.

We call the right component i of $\langle F, i\rangle$ *the depth of F in G*. We sometimes write $F \in Sub(G)$ instead of $\langle F, i\rangle \in Sub(G)$ if there is no confusion.

- Inference rules:

R1: Substitution

$$\frac{E == F \quad \langle E, i\rangle \in Sub(G)}{G == G(F/E)}.$$

R2: Replacement

$$\frac{E == F(e_1, \cdots, e_i, \cdots, e_n)}{E == F(e_1, \cdots, G/e_i, \cdots, e_n)}.$$

R3: Symmetry

$$\frac{E == F}{F == E}.$$

R4: Transitivity

$$\frac{E == F \quad F == G}{E == G}.$$

Here the common substitution operation sign (_/_) is used.

- Theorems of PA:

1. An axiom is a theorem.
2. If a formula $E == F$ is a theorem, and if we apply rules among R1–R4 to obtain a formula $E' == F'$, then $E' == F'$ is a theorem.
3. There are no exceptions.

Let us use the sign "$\vdash E == F$" to denote that $E == F$ is a theorem in PA, which is commonly used in contexts of formal derivation. Also for a set of formulas $\mathcal{S}$ as assumptions of derivation, let us write "$\mathcal{S} \vdash E == F$" as is commonly used as well.

6.3.2 Formal Results

Now let us observe various statements, all of which are given strict proof in Shimogawa and Takara (1994). Some of the important statements are attached with the proof. Other lemmas and propositions will be presented without the proof. Readers are guided to refer to Shimogawa and Takara (1994) in case they are interested in the technical threads.

Lemma 3.2

$$\vdash E == E.$$

Proof Omitted. ■

Lemma 3.3 *Let E, F, and $G \in Exp$ be arbitrary. Then,*

A: If $\vdash E == F$, then $\vdash (GE) == (GF)$.
B: If $\vdash E == F$, then $\vdash [E] == [F]$.

Proof Please refer to Shimogawa and Takara (1994). ■

Let us observe a notion of canonical form called *normal form* for the terms. It will be shown that every term can be represented by its "equivalent" normal form.

Definition 3.4 (Normal Form)

1. Variables and ε are normal forms.
2. $[E_i]$ is a normal form for any $E_i \in Exp$.
3. If E_i is a normal form, then $(E_i[E_j])$, $(E_i e_i)$, and $(E_i\varepsilon)$ are also normal forms, where E_j and e_i are arbitrary.
4. Nothing else is a normal form.

For an arbitrary $E \in Exp$, it is always possible to reform E into a normal form. In Shimogawa and Takara (1994), an algorithm is presented. (Please refer to Shimogawa and Takara (1994) for the procedure.)

Normal form is introduced simply for a technical reason. Roughly speaking a normal form of $E \in Exp$ is: (1) if $E = [F]$ for some $\langle F, j\rangle \in Sub(E)$, then E already is a normal form; (2) $E = ([E_1]\cdots[E_j])$, then re-arranging $\{[E_1], \cdots, [E_j]\}$ to the expression $((\cdots([E_{k_1}][E_{k_2}])[E_{k_3}])\cdots)[E_{k_i}])$ would be the normal form of E ($\{[E_1], \cdots, [E_j]\} = \{[E_{k_1}], \cdots, [E_{k_i}]\}$). For instance, if

$$E = (e_0([e_1]([(e_2[\varepsilon])][e_1]))),$$

then the normal form of E is

$$(([(e_2[\varepsilon])][e_1])e_0).$$

I introduce the notation $\overline{E}$ to represent the normal form of E.

Proposition 3.5 *For arbitrary expression $E \in Exp$,*

$$\vdash E == \overline{E}$$

holds.

Proof Please refer to Shimogawa and Takara (1994). ■

This fact implies that if we try to prove the syntactical statement such as $E == F$, we just have to verify $\overline{E} == \overline{F}$. Remember that "==" is transitive. Hereafter, in the syntactical arguments, it is sufficient to treat the normal form only and we will omit to write "(" and ")." That is, we write EF instead of (EF) for any $E, F \in Exp$. (In algebraic systems, this property is nothing other than *associativity* of binary operation (_, _).)

Lemma 3.6 *For arbitrary expression $E \in Exp$,*

$$\vdash [[E]] == E$$

holds.

Proof Let us see the demonstration of derivation as a typical inference of the system PA.

$$
\begin{array}{lll}
(a)\cdots & [[[E]][E]] == \varepsilon & (A1) \\
(b)\cdots & [[E]]\varepsilon == [[E]][[[E]][E]] & (a) \text{ and } (R1) \\
(c)\cdots & [[E]][[[E]][E]] == [[[E]][E]][[E]] & (A4) \\
(d)\cdots & [[E]]\varepsilon == [[E]]\varepsilon & (b), (c) \text{ and } (R4) \\
(e)\cdots & [[[E]][E]][[E]] == [[E]]\varepsilon & (d) \text{ and } (R3) \\
(f)\cdots & [[[E][[E]]][E[[E]]]] == [[[E]][E]][[E]] & (A2) \\
(g)\cdots & [[[E][[E]]][E[[E]]]] == [[E]]\varepsilon & (e), (f) \text{ and } (R4) \\
(h)\cdots & [[E]]\varepsilon == [[[E][[E]]][E[[E]]]] & (g) \text{ and } (R4) \\
(i)\cdots & [[E]]\varepsilon == [[[[E]][E]][[[E]]E]] & (h), (R4) \text{ and } (R1) \\
(j)\cdots & [[E]]\varepsilon == [\varepsilon[[[E]]E]] & (i), (R4) \text{ and } (R1) \\
(k)\cdots & [[E]]\varepsilon == [[[[E]]E]\varepsilon] & (j), (R4) \text{ and } (R1) \\
(l)\cdots & [[E]]\varepsilon == [[[[E]]E][[E]E]] & (j), (R4) \text{ and } (R4) \\
(m)\cdots & [[E]]\varepsilon == [[[[E]]][[E]]]E & (l), (R4) \text{ and } (A2) \\
(n)\cdots & [[E]]\varepsilon == \varepsilon E & (m), (R4) \text{ and } (R1) \\
 & [[E]] == E & (n) \text{ and } (R4). \blacksquare
\end{array}
$$

Corollary 3.7

$$
\begin{array}{ll}
(i) & \vdash EE == E, \\
(ii) & \vdash [E]E == [\varepsilon], \\
(iii) & \vdash E[FE] == E[F], \\
(iv) & \vdash E[\varepsilon] == [\varepsilon], \\
(v) & \vdash [E[F]][EF] == [E].
\end{array}
$$

Proof Almost trivial. Omitted. ■

6.3.3 Meta-Theorems: Soundness, Consistency, and Completeness

As we have already seen, in the original book of G.Spencer Brown, a term or an expression is interpreted as a representation of "state of indication." According to him, an expression can be simplified into the two symbols called "initials," which are the symbols of the situation "marked" and "unmarked," respectively. In the formulation I introduced, such a property of the original Primary Algebra can be realized as a two-valued interpretation function that is familiar in the context of logic.

Definition 3.8 (Formal Semantics of PA) Given a two-valued function $v : Exp \to \{0, 1\}$ satisfying:

$$(a) \quad v(e_i) = 1 \text{ or } 0, \text{ where } e_i \text{ is a variable.}$$

$$(b) \quad v([E]) = \begin{cases} 1 & \text{if } v(E) = 0 \\ 0 & \text{o.w.} \end{cases}$$

$$(c) \quad v((EF)) = \begin{cases} 1 & \text{if } v(E) = 1 \text{ or } v(F) = 1 \\ 0 & \text{o.w.} \end{cases}$$

$$(d) \quad v(\varepsilon) = 0,$$

we call the function v satisfying $(a) \sim (d)$ *an interpretation of PA*. Also,

$$\Sigma(PA) =_{def} \{v \mid v \text{ is an interpretation of } PA\}.$$

A formula $E == F$ in PA is called *tautology* if:

$$v(E) = v(F), \; \forall v \in \Sigma(PA).$$

To denote that $E == F$ is a tautology, the notation

$$\Sigma(PA) \models E == F$$

is employed, following common manner.

Definition 3.9 (Soundness) We say the system PA is *sound* if every theorem in PA is a tautology.

The next lemma allows us to omit "(" and ")" as well in the semantics.

Lemma 3.10 *For any $v \in \Sigma(PA)$, the following holds:*

$$v((E_1(E_2E_3)) = v(((E_1E_2)E_3)),$$

where E_1, E_2, and E_3 are terms of PA.

Proof Obvious. Omitted. (Please refer to Shimogawa and Takara (1994).) ■

Proposition 3.11

$$\Sigma(PA) \models E == \overline{E}.$$

Proof Due to Lemma 3.10. ■

At this point, we have seen the statements

$$\vdash E == \overline{E} \text{ and } \Sigma(PA) \models E == \overline{E}$$

are true. The implication is: We are ready to ignore binary operator (_, _). That is: We can write

$$E_1 \cdots E_i E_j \cdots E_k \cdots E_l \text{ instead of } (E_1, \ldots (\ldots (E_i, E_j)..,E_k)..,E_l)$$

Theorem 3.12 (Soundness) *System PA is sound.*

Proof It is done by induction on inference. Firstly we easily see (A1)∼ (A5) are tautologies. We will see Inference Rules (R1)∼(R4) preserve the property of tautology.

(R3),(R4): Trivial.

(R1): By induction along with the depth of subterms of arbitrary $G \in Exp$. Let $\langle E, i\rangle \in Sub(G)$. For the induction base, suppose $i = 1$. Then (R1) trivially preserves the property.
Assume any subterm of the depth n does not lose the property of tautology under application of (R1). Then for any subterm $\langle E', n+1\rangle \in Sub(G)$ there is a subterm $\langle F, n\rangle \in Sub(G)$ such that $F = [E'']$ and $\langle E'', n+1\rangle \in Sub(G)$, and $\langle E', 1\rangle \in Sub(E'')$. Let $\vdash E' == E'''$, then it yields

$$\vdash [E''] == [E''(E'''/E')]$$

from the fact $\langle E', 1\rangle \in Sub(E'')$ and Lemma 3.3. Since $F = [E'']$, the induction hypothesis yields

$$\Sigma(PA) \models G == G([E''(E'''/E')]/[E'']),$$

that is, $\Sigma(PA) \models G == G(E'''/E')$.

(R4): Obviously, since an arbitrary interpretation $v \in \Sigma(PA)$ is well defined as a function. ■

Definition 3.13 We say system PA is *consistent* if for any term $E \in Exp$ a formula $E == [E]$ can never be derived.

Theorem 3.14 (Consistency) *System PA is consistent.*

Proof Theorem 3.12 ensures that every formula derived in PA is a tautology. On the other hand, $E == [E]$ cannot be a tautology from the very definition of $\Sigma(PA)$. ■

Definition 3.15 (Completeness) System PA is said to be *complete* if:

$$\Sigma(PA) \models E == F \text{ iff } \vdash E == F$$

holds for arbitrary $E, F \in Exp$.

To see PA to be complete, several technical facts are in order.

Lemma 3.16 *Let $\mathcal{S}$ be a set of formulas. Then*

$$\text{If } \mathcal{S} \vdash E == F_1 \text{ and } \mathcal{S} \vdash E == F_2 \text{ then } \mathcal{S} \vdash E == F_1 F_2.$$

Omitted. Please refer to Shimogawa and Takara (1994). ■

Lemma 3.17 *For a set of formulas $\mathcal{S}$,*

$$\text{If } \mathcal{S} \vdash E == [\varepsilon] \text{ and } \mathcal{S} \vdash [E]F == [\varepsilon] \text{ then } \mathcal{S} \vdash F == [\varepsilon]$$

holds.

Proof We can easily see

$$\mathcal{S} \vdash E == [\varepsilon] \Leftrightarrow \mathcal{S} \vdash [E] == [[\varepsilon]] \Leftrightarrow \mathcal{S} \vdash [E] == \varepsilon,$$

hence applying (R1) to $[E]F == [E]F$ to obtain

$$\mathcal{S} \vdash [E]F == \varepsilon F,$$

therefore

$$\mathcal{S} \vdash [E]F == F,$$

and the assumption and (R4) yield $\mathcal{S} \vdash F == [\varepsilon]$. ■

Definition 3.18 For an expression $E \in Exp$, and an interpretation $v \in \Sigma(PA)$,

$$E^v =_{def} \begin{cases} E & \text{If } v(E) = 1 \\ [E] & o.w. \end{cases}$$

Lemma 3.19 *For $E \in Exp$, assume that all the variables occurring in E are $e_1, \cdots, e_j$. Then for arbitrary interpretation $v \in \Sigma(PA)$,*

$$\{e_1 == [\varepsilon], \cdots, e_j == [\varepsilon]\} \vdash E^v == [\varepsilon]$$

holds.

Proof See Shimogawa and Takara (1994). ■

Lemma 3.20 *Let $E, F \in Exp$ and $v \in \Sigma(PA)$ be arbitrary. Suppose all the variables occurring in E and F are $\{e_1, e_2, \cdots, e_j\}$. Then,*

$$\{e_1^v == [\varepsilon], \cdots, e_j^v == [\varepsilon]\} \vdash E^v == F^v.$$

The next statement corresponds to the well-known "deduction theorem" in the propositional calculus.

Corollary 3.21 *For terms E, F_1 and $F_2 \in Exp$ and a formula $F_1 == F_2$, suppose E contains no variables that occur in the formula $F_1 == F_2$. Then, if*

$$\mathcal{S} \cup \{E == [\varepsilon]\} \vdash F_1 == F_2,$$

we have

$$\mathcal{S} \vdash [E]F_1 == [E]F_2,$$

where $\mathcal{S}$ is a set of formulas in PA.

Proof Please refer to Shimogawa and Takara (1994) (somehow long). ■

Theorem 3.22 *System PA is complete.*

Proof We already had the only-if part, which is Theorem 3.12. Let us trace the derivation of $\vdash E == F$ from $\Sigma(PA) \models E == F$.

Our assumption: $\Sigma(PA) \models E == F$ for any $E, F \in Exp$. Lemma 3.20 tells us that for any $v \in \Sigma(PA)$ we have

$$\{e_1^v == [\varepsilon], \cdots, e_j^v == [\varepsilon]\} \vdash E^v == F^v. \tag{6.1}$$

From the assumption that $\Sigma(PA) \vdash E == F$, the formula $E^v == F^v$ has the form of $E == F$ or $[E] == [F]$. We can easily verify $\vdash E == F$ iff $\vdash [E] == [F]$ from Lemmas 3.3 and 3.6. Hence, it is sufficient to show that

$$\{e_1^v == [\varepsilon], \cdots, e_j^v == [\varepsilon]\} \vdash E == F \tag{6.2}$$

implies

$$\vdash E == F, \tag{6.3}$$

because the statement (6.2) is already given by Lemma 3.20 and the assumption. Therefore, proving the next is in order. For an arbitrary interpretation $v \in \Sigma(PA)$ and $i \in \{1, \cdots, j\}$,

$$\begin{aligned} &\text{If } \{e_1^v == [\varepsilon], \cdots, e_i^v == [\varepsilon]\} \vdash E == F, \\ &\text{then } \{e_1^v == [\varepsilon], \cdots, e_{i-1}^v == [\varepsilon]\} \vdash E == F. \end{aligned} \tag{6.4}$$

For the v we are able to select $v_1, v_2 \in \Sigma(PA)$ such that $e_k^{v_1} = e_k^{v_2} = e_k^v$ if $k \in \{1, \cdots, i-1\}$ and $v_1(e_k) \neq v_2(e_k)$ when $k = i$. For the v_1, v_2, Lemma 3.20 yields

$$v_1 : \{e_1^v == [\varepsilon], \cdots, e_{i-1}^v == [\varepsilon], e_i^{v_1} == [\varepsilon]\} \vdash E == F, \text{ and} \tag{6.5}$$

$$v_2 : \{e_1^v == [\varepsilon], \cdots, e_{i-1}^v == [\varepsilon], e_i^{v_2} == [\varepsilon]\} \vdash E == F. \tag{6.6}$$

Suppose $v_1(e_i) = 1$. Then $e_i^{v_1} = e_i$ and $e_i^{v_2} = [e_i]$. Hence,

$$v_1 : \{e_1^v == [\varepsilon], \cdots, e_{i-1}^v == [\varepsilon], e_i == [\varepsilon]\} \vdash E == F, \text{ and} \tag{6.7}$$

$$v_2 : \{e_1^v == [\varepsilon], \cdots, e_{i-1}^v == [\varepsilon], [e_i] == [\varepsilon]\} \vdash E == F. \tag{6.8}$$

We apply Corollary 3.21 to (6.7) and (6.8) to obtain

$$(6.7) \Rightarrow \{e_1^v == [\varepsilon], \cdots, e_{i-1}^v == [\varepsilon]\} \vdash [e_i]E == [e_i]F \text{ and} \tag{6.9}$$

$$(6.8) \Rightarrow \{e_1^v == [\varepsilon], \cdots, e_{i-1}^v == [\varepsilon]\} \vdash [[e_i]]E == [[e_i]]F. \tag{6.10}$$

From (6.10) we immediately obtain

$$\{e_1^v == [\varepsilon], \cdots, e_{i-1}^v == [\varepsilon]\} \vdash e_i E == e_i F \tag{6.11}$$

using Lemma 3.6. The relations (6.9) and (6.11) yield the following:

$$(6.9) \Rightarrow \{e_1^v == [\varepsilon], \cdots, e_{i-1}^v == [\varepsilon]\} \vdash [[e_i]E] == [[e_i]F] \text{ and} \tag{6.12}$$

$$(6.11) \Rightarrow \{e_1^v == [\varepsilon], \cdots, e_{i-1}^v == [\varepsilon]\} \vdash [e_i E] == [e_i F]. \tag{6.13}$$

Then, Lemma 3.16 implies

$$\{e_1^v == [\varepsilon], \cdots, e_{i-1}^v == [\varepsilon]\} \vdash [[e_i]E][e_i E] == [[e_i]F][e_i F]. \tag{6.14}$$

Applying Corollary 3.7 to $[[e_i]E][e_i E]$ and $[[e_i]F][e_i F]$, we obtain

$$\vdash [[e_i]E][e_i E] == [E] \text{ and } \vdash [[e_i]F][e_i F] == [F], \tag{6.15}$$

respectively. Then (6.14) yields

$$\{e_1^v == [\varepsilon], \cdots, e_{i-1}^v == [\varepsilon]\} \vdash [E] == [F],$$

which immediately implies

$$\{e_1^v == [\varepsilon], \cdots, e_{i-1}^v == [\varepsilon]\} \vdash E == F.$$

Consequently, (6.4) holds and hence we have (6.3). ■

6.3.4 Reformation of Propositional Calculus

In order to establish a natural mapping from each term in PA to wff of propositional calculus, we review and slightly reform the usual syntax of propositional calculus as follows. Let us call this inference system $PC*$.

Definition 3.23 (System $PC*$**)** Language of $PC*$: $L(PC*)$

1. Primitive symbols of $L(PC*)$
   - Variables: $\{p_i\}_{i \in \mathbb{N}}$
   - False symbol: $\{F\}$
   - Connectives: $\{\vee, \neg\}$
   - Brackets: {[,]}
2. Wffs of $L(PC*)$: $Form(L(PC*))$
   - Variables and F are wffs.
   - If A is a wff, then $\neg A$ also is.
   - If A and B are wffs, then $[A \vee B]$ is also a wff.
   - Nothing else is wff.

Axioms: Let P, Q, R be all wffs of L(PC*).

A1 $[P \vee P] \rightarrow P$,
A2 $P \rightarrow [P \vee Q]$,
A3 $[P \vee Q] \rightarrow [Q \vee P]$,
A4 $[P \rightarrow Q] \rightarrow [[R \vee P] \rightarrow [R \vee Q]]$,
A5 $F \equiv \neg[P \vee \neg P]$,

where $[P \rightarrow Q]$ is an abbreviation for $[\neg P \vee Q]$, $P \equiv Q$ for $[P \rightarrow Q] \wedge [Q \rightarrow P]$ and $[R \wedge S]$ abbreviates $\neg[\neg R \vee \neg S]$, as are conventional. Notice that "[,]" are frequently omitted in case of no confusion.

Inference Rule: (Modus Ponens)

$$\frac{A \quad [A \rightarrow B]}{B}\text{MP.}$$

We use the conventional notation "$\vdash$" also for the system $PC*$, that is, we write

$$PC* \vdash P$$

if a wff P in $L(PC*)$ is derived in the inference system.

The semantics of $PC*$ is also defined in a usual way.

Definition 3.24 (Formal Semantics of $PC*$**:** $\Sigma(PC*)$**)** A Boolean valued mapping

$$v : Form(L(PC*)) \rightarrow \{0, 1\}$$

satisfying:

- $v(p_i) = 0$ or 1 for all $p_i \in \{p_i\}$ (variables of $L(PC*)$),
- $v(\neg P) = 1$ iff $v(P) = 0$,
- $v(P \vee Q) = 0$ iff $(v(p) = 0$ and $v(q) = 0)$,
- $v(F) = 0$,

is called *an interpretation of* $PC*$. Also, let $\Sigma(PC*)$ denote the set of all the interpretations of $PC*$:

$$\Sigma(PC*) = \{v \mid v : \text{ an interpretation of } PC*\}.$$

Of course, the definitions above are slight reform of standard propositional calculus (without any relevant extension), and we could omit to review and show its main characteristics, meaning "completeness" as is well-known.

6.3.5 A "Natural" Mapping from PA to $PC*$

Now we are ready to proceed to construct a correspondence between the system PA and $PC*$. It is done with wffs of both systems, and later on we will see a "consistency" of the correspondence.

Definition 3.25 A function

$$\tau : Exp \to Form(L(PC*))$$

is defined recursively as

- $\tau(e_i) = p_i$ for $i \in \mathbb{N}$,
- $\tau(\varepsilon) = F$,
- $\tau([E]) = \neg\tau(E)$,
- $\tau(EG) = \tau(E) \vee \tau(G)$.

Of course the function τ has some good characteristics such as

Lemma 3.26 *τ is bijective.*

Proof Just like τ defined above, let me introduce the map $\sigma : Form(L(PC*)) \to Exp$ as

- $\sigma(p_i) = e_i$ for $i \in \mathbb{N}$,
- $\sigma(F) = \varepsilon$,
- $\sigma(\neg P) = [\sigma(P)]$,
- $\sigma(P \vee Q) = \sigma(P)\sigma(Q)$.

Claim:

$$\sigma(\tau(E)) = E, \ \forall E \in Exp. \tag{6.16}$$

To show (6.16), it is enough to see a simple induction along with the construction of E.

(Base:) Trivial.

(Steps:) Let the induction hypothesis be $\sigma(\tau(E)) = E$ and $\sigma(\tau(F)) = F$. Then

$$\begin{aligned}\sigma(\tau([E])) &= \sigma(\neg\tau(E)) \\ &= [\sigma(\tau(E))] \\ &= [E], \text{ and}\end{aligned}$$

$$\begin{aligned}\sigma(\tau(EF)) &= \sigma(\tau(E) \vee \tau(F)) \\ &= \sigma(\tau(E))\sigma(\tau(F)) \\ &= EF.\end{aligned}$$

Claim:

$$\tau(\sigma(P)) = P, \ \forall P \in Form(L(PC*)). \tag{6.17}$$

The proof is omitted, for it is almost the same way as was used proving (6.16).

Now we have obtained a couple of bijective maps τ and its inverse $\sigma = \tau^{-1}$ (as showing (6.16) and (6.17)), the conclusion is ensured. ■

Lemma 3.27

$$\tau^{-1}(\tau(E) \equiv \tau(F)) = [[[E]F][[F]E]].$$

Proof

$$\begin{aligned}\tau^{-1}(\tau(E) \equiv \tau(F)) &= \sigma(\tau(E) \equiv \tau(F)) && \because \tau^{-1} = \sigma \\ &= \sigma((\neg\tau(E) \vee \tau(F)) \wedge (\neg\tau(F) \vee \tau(E))) && \because \text{definition of } \equiv \\ &= \sigma(\neg(\neg(\neg\tau(E) \vee \tau(F)) \vee \neg(\neg\tau(F) \vee \tau(E)))) && \because \text{definition of } \wedge \\ &= \sigma(\neg(\neg(\tau([E]) \vee \tau(F))) \vee \neg(\tau([F]) \vee \tau(E))) && \because \text{definition of } \tau \\ &= \sigma(\neg(\neg(\tau([E]F))) \vee \neg(\tau([F]E))) && \because \text{definition of } \tau \\ &= \sigma(\neg(\tau([[E]F]) \vee \tau([[F]E]))) && \because \text{definition of } \tau \\ &= \sigma(\neg\tau([[E]F][[F]E])) && \because \text{definition of } \tau \\ &= [\sigma(\tau([[E]F][[F]E]))] && \because \text{definition of } \sigma \\ &= [[[E]F][[F]E]] && \because \sigma(\tau(E)) = E.\end{aligned}$$ ■

Lemma 3.28

$$\Sigma(PA) \models [[[E]F][[F]E]] == [\varepsilon] \quad \textit{if and only if} \quad \Sigma(PA) \models F == E.$$

Proof Omitted. It is based on the definition of v: the interpretation of PA (Please refer to Shimogawa and Takara (1994)). ■

Lemma 3.29 *If* $\vdash E == F$*, then*

$$\Sigma(PC*) \models \tau([[[E]F][[F]E]]).$$

Proof By induction along with the inference process.

(Bases:) If formula $E == F$ is one of the axioms of PA, it is simple and clear. I will provide here the reasoning in case of A2: $E = [[E_1E_3][E_2E_3]]$ and $F = [[E_1][E_2]]E_3$. From Lemma (3.27),

$$\tau(E) \equiv \tau(F) = \tau([[[E]F][[F]E]])$$

holds. Hence it is sufficient to show

$$v([[E_1E_3][E_2E_3]]) = v([[E_1][E_2]]E_3), \quad \forall v \in \Sigma(PC*).$$

The calculation goes as follows:

$$\begin{aligned}
\tau([[E_1E_3][E_2E_3]]) &= \neg(\tau([E_1E_3]) \vee \tau([E_2E_3])) \\
&= \neg(\neg(\tau(E_1) \vee \tau(E_3)) \vee \neg(\tau(E_2) \vee \tau(E_3))) \\
&\equiv (\tau(E_1) \vee \tau(E_3)) \wedge (\tau(E_2) \vee \tau(E_3)) \\
&\equiv (\tau(E_1) \wedge \tau(E_2)) \vee \tau(E_3) \\
&\equiv \neg(\neg\tau(E_1) \vee \neg\tau(E_2)) \vee \tau(E_3) \\
&= \tau([[E_1][E_2]]E_3),
\end{aligned}$$

therefore $\tau([[E_1E_3][E_2E_3]]) \equiv \tau([[E_1][E_2]]E_3)$, and the completeness of the system $PC*$ yields

$$v(\tau([[E_1E_3][E_2E_3]])) = v(\tau([[E_1][E_2]]E_3)), \quad \forall v \in \Sigma(PC*).$$

(Steps:) Among inference rules of PA, R1 and R2 seem to be essential. R2 case is quite similar to R1 case, so I provide only R1 case here.

The goal is

$$\begin{aligned}
&\text{If } \Sigma(PC*) \models \tau([[[E]F][[F]E]]), \text{ then} \\
&\quad \Sigma(PC*) \models \tau([[[G]G(F/E)][[G(F/E)]G]]),
\end{aligned}$$

for, from Lemma 3.27, $\tau([[[E]F][[F]E]]) = \tau(E) \equiv \tau(F)$ always holds. In this case the completeness of $PC*$ implies $v(\tau(E)) = v(\tau(F))$ for all $v \in \Sigma(PC*)$. From this fact it is sufficient to show that

$$v(\tau(G)) = v(\tau(G(F/E))),$$

and then we obtain

$$v(\tau([[[G]G(F/E)][[G(F/E)]G]])) = 1$$

using Lemma 3.27 again.

In accordance with the definition of τ we can easily see that $\tau(G(F/E)) = \tau(G)(\tau(F)/\tau(E))$. Also it is obvious that

$$\text{If } v(P) = v(Q) \text{ and } P \in Sub(R), \text{ then } v(R) = v(R(Q/P))$$

in propositional calculus. Hence, our assumption $v(\tau(E)) = v(\tau(F))$ yields $v(\tau(G)) = v(\tau(G)(\tau(F)/\tau(E))) = v(\tau(G(F/E)))$, which is the result we wanted. ■

Lemma 3.30

If $PC* \vdash \tau([[[E]F][[F]E]])$, *then* $\Sigma(PA) \models [[[E]F][[F]E]] == [\varepsilon]$.

Proof Its derivation is straightforward. I omit the proof. Please refer to Shimogawa and Takara (1994). ■

6.3.6 *Main Statements and Implications*

6.3.6.1 Theorems

Theorem 3.31

$$\Sigma(PA) \models E == F \tag{6.18}$$

if and only if

$$\Sigma(PC*) \models \tau(E) \equiv \tau(F). \tag{6.19}$$

Proof Suppose (6.18) holds. From the completeness of PA (Theorem 3.22), we have

$$PA \vdash E == F.$$

Then Lemma 3.29 yields

$$\Sigma(PC*) \models \tau([[[E]F][[F]E]]).$$

From Lemma 3.27 we already know

$$\tau([[[E]F][[F]E]]) = \tau(E) \equiv \tau(F),$$

which completes "only if" part. On the other hand, using Lemma 3.27 again and assumption (6.19) we have

$$\Sigma(PC*) \models \tau([[[E]F][[F]E]]),$$

which of course yields

$$PC* \vdash \tau([[[E]F][[F]E]])$$

from the completeness of system $PC*$, and again Lemmas 3.28 and 3.30 justifies

$$\Sigma(PA) \models E == F. \blacksquare$$

Theorem 3.32

$$PA \vdash E == F \tag{6.20}$$

if and only if

$$PC* \vdash \tau(E) \equiv \tau(F). \tag{6.21}$$

Proof From the completeness of PA and $PC*$, and Theorem 3.31, we have

$$PA \vdash \sigma(\tau(E) \equiv \tau(F)) == [\varepsilon] \text{ if and only if } PC* \vdash \tau(\sigma(\tau(E) \equiv \tau(F))).$$

And Lemma 3.26 implies

$$PA \vdash \sigma(\tau(E) \equiv \tau(F)) == [\varepsilon] \text{ if and only if } PC* \vdash \tau(E) \equiv \tau(F).$$

Hence it is sufficient to show

$$PA \vdash \sigma(\tau(E) \equiv \tau(F)) == [\varepsilon] \text{ if and only if } PA \vdash E == F. \tag{6.22}$$

From the completeness of PA, the statement (6.22) is equivalent to

$$\Sigma(PA) \models \sigma(\tau(E) \equiv \tau(F)) == [\varepsilon] \text{ if and only if } \Sigma(PA) \models E == F. \tag{6.23}$$

$\sigma(\tau(E) \equiv \tau(F))$ is equal to $[[[E]F][[F]E]]$ (see the definition of "$\equiv$" and Lemma 3.27). Hence, Lemma 3.28 can be directly applied and we finish the proof. ■

6.3.6.2 Implications

At this point it could be worthwhile noting that the Definition 3.25:

$$\tau : Exp \rightarrow Form(L(PC*)),$$

its inverse:

$$\sigma : Form(L(PC*)) \rightarrow Exp$$

and the "semantics" of "==" in PA seem to have an essential role in the story. These notions "bridge" between "known framework" and "unknown/ambiguous notions." And at least, due to the discussion demonstrated above (Shimogawa and Takara, 1994), it is appropriate to say that the system PA can be embedded into the propositional calculus.

6.4 Story 2: Reformation of Brownian Algebra as an Algebraic System

So, let us begin with the "second" story, the start point of which would be the arrow (b) in the diagram presented in the introductory section.

This section consists of the discussions based upon mainly the following two viewpoints: (1) algebraic adjustment of F. J. Varela's developments of the notion "Primary Algebra," which we discussed in previous sections. The developments done by F.J.Varela are based on the book "PRINCIPLES OF BIOLOGICAL AUTONOMY" (Varela, 1979). The story itself is totally based on Shimogawa (1998).

In the first stage, the author set up the framework borrowed from the notion of universal algebra (Graetzer, 1979) in order to transform "Primary Algebra" and its extension "Brownian Algebra" to algebraic systems.

6.4.1 Basic Notions

An *algebra* is a pair $\langle A; \mathcal{F}\rangle$, where A is a set and $\mathcal{F}$ is a class of functions called the *operators*. An operator f of $\langle A; \mathcal{F}\rangle$ is adjoined with its *domain* $dom(f) = A^{\gamma}$ for some ordinal γ and codomain $cod(f) : A$. An operator $f \in \mathcal{F}$ with $dom(f) = A^\gamma$ is a set of ordered pair:

$$\{((\cdots, a_\alpha, \cdots), a) \mid (\cdots, a_\alpha \cdots)_{\alpha<\gamma} \in A^\gamma \;\&\; a \in A\}$$

in which $((\cdots, a_\alpha, \cdots), a) \in f$ and $((\cdots, a_\alpha, \cdots), b) \in f$ implies $a = b$. An operator is written in $f : A^\gamma \to A$, as is conventional. Also the *composition of the operators* is written following common notation: $f(- - -.f_\gamma, - - -)$ and so on. It is always assumed that a symbol never represents two or more functions, and all elements of $\mathcal{F}$ have a unique name in the algebra, that is:

Definition 4.1 (Denotation) Given a class $\overline{\mathcal{A}}$ of algebras $(= \{\cdots, \langle A; \mathcal{F}_A\rangle, \cdots\},)$ and a set $\overline{F}$ of symbols (assuming denumerable), a correspondence $Denot \subset \overline{F} \times \overline{\mathcal{F}}$ (where $\overline{\mathcal{F}}$ denotes the set of all operators of algebras in $\overline{\mathcal{A}}$) such that,

for each algebra $\langle A; \mathcal{F}_A\rangle \in \overline{\mathcal{A}}$, there exists $F \subset \overline{F}$ which satisfies that $Denot(F) = \mathcal{F}_A$ and $Denot_F \subset F \times \mathcal{F}_A$ is a bijective mapping,

is called the *denotation*. Let us call such F the operator symbols for $\mathcal{F}_A$.

In the definition above, we assume the following conventional notation:

For $E \subseteq A \times B$,

$$E(A) =_{def} \{b \in B \mid (a, b) \in E \text{ for some } a \in A\},$$

and for some $C \subseteq A$,

$$E_C =_{def} \{(a, b) \in E \mid a \in C\}.$$

Let us use the notation $\langle A; F\rangle$ to represent $\langle A; \mathcal{F}\rangle$. If an algebra $\langle A; F\rangle$ is given, we use Sans Serif character f, g, and so on, to represent the element of F.

Definition 4.2 Given an algebra $\langle A; F\rangle$, let $\mathsf{f} \in F$ and all of $\cdots, a_\gamma, \cdots, a \in A$. We write

$$f(\cdots, a_\gamma, \cdots) = a$$

to represent

$$((\cdots, a_\gamma, \cdots), a) \in Denot(\{\mathsf{f}\}).$$

Definition 4.3 (Types) Given an algebra $\langle A, F\rangle$, the *type of the algebra* is a function $\mathtt{t} : F \to \omega$, which satisfies

$$\mathtt{t} : F \ni \mathtt{f} \mapsto \gamma \in \omega \text{ if and only if } dom(f) = A^\gamma.$$

An algebra $\langle A;\, F\rangle$ is said to *have a subtype of* an algebra $\langle B;\, G\rangle$ if there exists an injective mapping $\phi : F \to G$ with the property

$$\mathtt{t}_F = \mathtt{t}_G \circ \phi,$$

where $\mathtt{t}_F$ and $\mathtt{t}_G$ represent types of $\langle A;\, F\rangle$ and $\langle B;\, G\rangle$, respectively, and "$\circ$" is conventional notation of composition of two functions.

A pair of algebras is said to *have the same type* if each of them has the subtype of another.

A class of algebras $\{\langle A_j;\, F_j\rangle\}_{j\in J}$ is said to have the same type if for all pair k, l in J, $\langle A_k;\, F_k\rangle$ and $\langle A_l;\, F_l\rangle$ have the same types.

Definition 4.4 (Polynomials) Given a set of variables $Ver = \{\cdots, \mathtt{x}, \cdots\}$ in our language (the domain of them is underlying set of given algebra), let an algebra $\langle A;\, F\rangle$, $F = \{\cdots, \mathtt{f}_\alpha, \cdots\}_{\alpha<\gamma}$, α and γ are some ordinals, be given. *A polynomial* of the algebra is defined as

- All $\mathtt{x} \in Ver$ is a polynomial of the algebra.
- If $\{\cdots, \mathtt{p}_\beta, \cdots\}_{\beta\leq\sigma}$ are polynomials of the algebra (β and σ are ordinals), then $\mathtt{f}(\cdots, \mathtt{p}_\beta, \cdots)$ is a polynomial for any $\mathtt{f} \in F$ with $\mathtt{t}(\mathtt{f}) = \sigma$.
- Nothing else is a polynomial of the algebra.

If there exists a familiar notation for an operator symbol of a particular operation in some algebra, we will adopt it. Namely, we write $x \vee y$ instead of $\vee(x, y)$ and so on. We adopt the notation T_F to represent the set of all polynomials of the algebra $\langle A;\, F\rangle$. We say a polynomial $\mathtt{p}$ is constituted of $\Phi = \{\cdots, \mathtt{f}, \cdots\} \subseteq F$ if an operator symbol $\mathtt{f}$ occurs in $\mathtt{p}$ if and only if $\mathtt{f} \in \Phi$.

The *arity* of a polynomial $\mathtt{p}$ is the cardinality of occurrences of variables in $\mathtt{p}$. We call $\mathtt{p}$ a γ-ary polynomial if its arity is γ. For example, $\mathtt{f(x, g(y, x), z)}$ is a 3-ary polynomial, whereas $\mathtt{f(x, g(y, w), z)}$ is a 4-ary. For each arity γ of polynomials of the given algebra, let us denote the set of γ-ary polynomials by T_F^γ. If $\mathtt{p}$ and $\mathtt{q}$ are polynomials, then $\mathtt{p} = \mathtt{q}$ is called *an equation of the algebra*.

The "denotation" of a polynomial can be defined to be composition of functions all of which are denotations of operator symbols occurring in the polynomial. Thus we can naturally extend the notion of $Denot_F$ to construct the new mapping $\overline{Denot}_F$, whose domain is the set of all polynomials of given algebra.

Definition 4.5 (States) Let an algebra $\langle A;\, F\rangle$ be given. *A state* of the algebra is a mapping

$$s^A : Ver \to A.$$

Σ_A is used to represent the set of all states of the algebra.

Let us consider how to construct proper "interpretation" of polynomials on a particular algebra. To do that, it is convenient to regard a variable $\mathtt{x} \in Var$ as the mapping that maps a state s^A to an element $s^A(\mathtt{x})$ of base set of the algebra. Formally:

$$\begin{aligned} \overline{\mathtt{x}} : \Sigma_A &\rightarrow A \\ : s^A &\mapsto s^A(\mathtt{x}). \end{aligned}$$

Viewing $\mathtt{x}$ as the mapping $\overline{\mathtt{x}}$, the interpretation of a polynomial $\mathtt{f}(\mathtt{x}, \mathtt{y})$ on the algebra $\langle A; F\rangle$, for instance, can be viewed as

$$\begin{aligned} \mathtt{f}(\overline{\mathtt{x}}, \overline{\mathtt{y}}) : \Sigma_A &\rightarrow A \\ : s^A &\mapsto f(s^A(\mathtt{x}), s^A(\mathtt{y})) \end{aligned}$$

Definition 4.6 (Denotation of Polynomials) Given an algebra $\langle A; F\rangle$, the mapping $\overline{Denot}_F$ on all polynomials of the algebra is defined as

- If $\mathtt{x} \in Ver$, then $\overline{Denot}_F(\mathtt{x}) = \overline{\mathtt{x}}$,
- If $\mathtt{f} \in F$, $\mathtt{t}(\mathtt{f}) = \gamma$ and $\{\cdots, \mathtt{p}_\alpha, \cdots\}_{\alpha \leq \gamma} \subset T^F$, each of whose denotation is defined, then $\overline{Denot}_F(\mathtt{f}(\cdots, \mathtt{p}_\alpha, \cdots)) = f(\cdots, \overline{Denot}_F(\mathtt{p}_\alpha), \cdots)$, as is defined as composition of functions.

Definition 4.7 (Validity of an Equation) Given an algebra $\langle A; F\rangle$, an equation $\mathtt{p} = \mathtt{q}$ is said to be *satisfied* by the state s^A $(\in \Sigma_A)$ provided

$$\overline{Denot}_F(\mathtt{p})(s^A) = \overline{Denot}_F(\mathtt{q})(s^A)$$

holds. If

$$\overline{Denot}_F(\mathtt{p})(s^A) = \overline{Denot}_F(\mathtt{q})(s^A), \quad \forall s^A \in \Sigma_A$$

holds, then we say $\mathtt{p} = \mathtt{q}$ is *valid* in the algebra.

For any $\mathtt{p}, \mathtt{q} \in T_F$, define a binary relation $\equiv \subset T_F \times T_F$ as

$$\mathtt{p} \equiv \mathtt{q} \text{ iff } \mathtt{p} = \mathtt{q} \text{ is valid in } \langle A; F\rangle.$$

"$\equiv$" is an equivalence relation on T_F and the notation $[\mathtt{q}]$ is adopted to denote elements of quotient class on T_F.

Let us use notation $\mathtt{p}(\mathtt{r}/\mathtt{q})$ to represent the resulting polynomial of substitution $\mathtt{r}$ for $\mathtt{q}$ in $\mathtt{p}$.

Similarly $(\mathtt{p} = \mathtt{q})(\mathtt{r}/\mathtt{s})$ denotes the polynomial, which is the result of substitution of $\mathtt{r}$ for $\mathtt{s}$ in $\mathtt{p}$ and $\mathtt{q}$.

FACT 4.8

(1) If an equation $\mathtt{p} = \mathtt{q}$ *is valid in the given algebra and has an occurrence of a variable* $\mathtt{x} \in Var$*, then for an arbitrary polynomial* $\mathtt{u}$*,*

$$(\mathtt{p} = \mathtt{q})(\mathtt{u}/\mathtt{x})$$

also is valid.

(2) If an equation $\mathtt{p} = \mathtt{q}$ *is valid in the algebra, then for any polynomial* $\mathtt{u}$ *in which variable* $\mathtt{x} \in Var$ *occurs,*

$$(\mathtt{u})(\mathtt{p}/\mathtt{x}) = (\mathtt{u})(\mathtt{q}/\mathtt{x})$$

is also a valid equation.

Proof Obvious. ■

If we consider a system model as an algebra, its external expression should be written by "axioms" rather than building up the operators' correspondence table. Axioms of an algebraic system are to consist of "equations," which manner is quite normal in the usual mathematical context. Also Fact 4.8 is the tool of the derivations to find out characteristics of a given algebraic system. Practically, we use Fact 4.8 without any remark.

Definition 4.9 (Consistency) An algebra $\langle A; F\rangle$ is said to be *consistent* if there exist equations that can never be derived using Fact 4.8. If not, the algebra is said to be *inconsistent* or *trivial.*

Of course we are not going to talk about trivial algebraic systems.

As explained earlier, a class of algebraic systems of "the same property" is to be identified by a common set of "axioms."

FACT 4.10 *If an algebra* $\langle A; F\rangle$ *is inconsistent, then the underlying set* A *is a singleton set.*

Proof Every equation on the algebra is valid, hence for any $s^A \in \Sigma_A$,

$$\overline{Denot}_F(\mathtt{x})(s^A) = \overline{Denot}_F(\mathtt{y})(s^A)$$

for any variables $\mathtt{x}, \mathtt{y} \in Var$. So for any $a, b \in A$, $a = b$ is derived. ■

If an algebra $\langle A; F\rangle$ with its type $\mathtt{t}_F$ has subtype of an algebra $\langle B; G\rangle$ with type $\mathtt{t}_G$, and $\mathtt{t}_F = \mathtt{t}_G \circ \phi$, then we have the injective mapping $\hat{\phi} : T_F \to T_G$ induced by ϕ, following usual induction.

$$\hat{\phi} : \mathtt{x} \mapsto \mathtt{x} \text{ if } \mathtt{x} \in Var,$$

$$: \mathtt{f}(\cdots, \mathtt{p}, \cdots) \mapsto \phi(\mathtt{f})(\cdots, \hat{\phi}(\mathtt{p}), \cdots), \quad \mathtt{p} \in T_F \text{ and } \mathtt{f} \in F.$$

Definition 4.11 (Weakness) Let a pair of algebras $\langle A;\ F\rangle$ and $\langle B;\ G\rangle$ be given. Let $\mathtt{t}_F$ and $\mathtt{t}_G$ be their types, respectively. We say the algebra $\langle A;\ F\rangle$ is a *weak algebra* of $\langle B;\ G\rangle$ provided:

- $\langle A;\ F\rangle$ has the subtype of $\langle B;\ G\rangle$, (let $\mathtt{t}_F = \mathtt{t}_G \circ \phi$).
- If $\mathtt{p} = \mathtt{q}$ is valid in $\langle A;\ F\rangle$, then so is $\hat{\phi}(\mathtt{p}) = \hat{\phi}(\mathtt{q})$ in $\langle B;\ G\rangle$.

6.4.2 Formalizing Primary Algebra and Brownian Algebra

The set of variables Var is always supposed to be given. As was discussed earlier, we use a set of equations to "characterize" an algebraic system.

Definition 4.12 (Primary Algebra (Spencer Brown, 1969)) Suppose $\mathtt{x}, \mathtt{y}, \mathtt{z} \in Var$. A primary algebra is an algebra $\langle A;\ F\rangle$ where $F = \{(\_,\_), [\_], \varepsilon\}$, $\mathtt{t}((\_,\_)) = 2$, $([\_]) = 1$, $\mathtt{t}(\varepsilon) = 0$ in which the following equations are valid:

$$\begin{aligned} P1: &\quad (\mathtt{xy}) = (\mathtt{yx}),\\ P2: &\quad ((\mathtt{xy})\mathtt{z}) = (\mathtt{x}(\mathtt{yz})),\\ P3: &\quad [([\mathtt{x}]\mathtt{x})] = \varepsilon,\\ P4: &\quad [([(\mathtt{xz})][(\mathtt{yz})])] = ([([\mathtt{x}][\mathtt{y}])]\mathtt{z}),\\ P5: &\quad (\varepsilon\mathtt{x}) = \mathtt{x}. \end{aligned}$$

Note that the nullary operator symbol ε represents a particular element in A. We call $P1 \cdots P5$ *axioms* of primary algebra.

The operator symbol [_] corresponds to "$\rceil$" in the original notation (Spencer Brown, 1969), and was also used in the story 1. Also, $P2$ is the associativity of the 2-ary operation (_, _), so in the following discussion the operator symbols (and) are sometimes omitted.

Let us use the abbreviation PA to represent primary algebra, if the context is not confusing with story 1.

In Spencer Brown (1969) and Varela (1979) a number of demonstrations of "derivations" were given. In this context, the "derivations" are to be examined and performed again under the framework we are observing in this section.

F.J. Varela introduced the notion of "Brownian Algebra" in order to describe "wave form calculus" (Varela, 1979). Brownian Algebra is also formalized with our framework as follows.

Definition 4.13 (Brownian Algebra) A *Brownian algebra* (BA will be used for the abbreviation) is an algebra $\langle B; G \rangle$ where $G = \{(\_, \_), [\_], \varepsilon\}$, the type of which is the same as that of PA. Following equations are always valid in BA, namely, they are the axioms of BA.

$$\begin{aligned} B1: &\quad (\mathtt{xy}) = (\mathtt{yx}), \\ B2: &\quad ((\mathtt{xy})\mathtt{z}) = (\mathtt{x}(\mathtt{yz})), \\ B3: &\quad ([([\mathtt{x}]\mathtt{y})]\mathtt{x}) = \mathtt{x}, \\ B4: &\quad [([(\mathtt{xz})][(\mathtt{yz})])] = ([([\mathtt{x}][\mathtt{y}])]\mathtt{z}), \\ B5: &\quad (\varepsilon\mathtt{x}) = \mathtt{x}. \end{aligned}$$

The only one difference between PA and BA is $P3 - B3$.

FACT 4.14 *PA and BA are both consistent.*

Proof We can easily set up the operations that satisfy $P1..P5$ and also $B1 \ldots B5$ with the base set $\{0, 1\}$. ■

FACT 4.15 *In PA (and BA), if there is another nullary operator ε' satisfying*

$$B5': \quad (\varepsilon'\mathtt{x}) = \mathtt{x},$$

then $\varepsilon = \varepsilon'$ is always valid.

Proof

$$\begin{aligned} &(a) \quad \mathtt{x} = (\varepsilon\mathtt{x}) && (\because P5) \\ &(b) \quad \varepsilon' = (\varepsilon\varepsilon') && (\because \text{fact4.8}) \\ &(c) \quad (\varepsilon'\mathtt{x}) = \mathtt{x} && (\because \text{assumption}) \\ &(d) \quad (\varepsilon'\varepsilon) = \varepsilon && (\because \text{fact4.8}) \\ &(e) \quad (\varepsilon'\varepsilon) = (\varepsilon\varepsilon') && (\because P1) \\ &(f) \quad \varepsilon' = \varepsilon && (\because (b), (d), (e)) \blacksquare \end{aligned}$$

So, we only need a single nullary operator symbol in both PA and BA.
Let us introduce notations:

$$\Sigma_{PA} =_{def} \{P1, \ldots P5\} \text{ and}$$

$$\Sigma_{BA} =_{def} \{B1, \ldots B5\}.$$

They are called the *axioms of* PA or BA, respectively.

It would be a good idea for us to review axioms of "Boolean algebra" at this point. An algebra $\langle A; \{\vee, \wedge\}\rangle$ with the type: $\mathtt{t}(\vee) = \mathtt{t}(\wedge) = 2$ is called a *lattice* if following equations are valid:

$$\begin{array}{lll} L1: & \mathtt{x} \vee \mathtt{x} = \mathtt{x}, & \mathtt{x} \wedge \mathtt{x} = \mathtt{x}, \\ L2: & \mathtt{x} \vee \mathtt{y} = \mathtt{y} \vee \mathtt{x}, & \mathtt{x} \wedge \mathtt{y} = \mathtt{y} \wedge \mathtt{x}, \\ L3: & \mathtt{x} \vee (\mathtt{y} \vee \mathtt{z}) = (\mathtt{x} \vee \mathtt{y}) \vee \mathtt{z}, & \mathtt{x} \wedge (\mathtt{y} \wedge \mathtt{z}) = (\mathtt{x} \wedge \mathtt{y}) \wedge \mathtt{z}, \\ L4: & \mathtt{x} \wedge (\mathtt{x} \vee \mathtt{y}) = \mathtt{x}, & \mathtt{x} \vee (\mathtt{x} \wedge \mathtt{y}) = \mathtt{x}. \end{array}$$

A *distributive lattice* is a lattice in which the equation of distributivity holds

$$L5: \quad \mathtt{x} \vee (\mathtt{y} \wedge \mathtt{z}) = (\mathtt{x} \vee \mathtt{y}) \wedge (\mathtt{x} \vee \mathtt{z})$$

FACT 4.16 *In a distributive lattice, the following is always valid:*

$$L5': \quad \mathtt{x} \wedge (\mathtt{y} \vee \mathtt{z}) = (\mathtt{x} \wedge \mathtt{y}) \vee (\mathtt{x} \wedge \mathtt{z}).$$

Proof Note that we always use Fact 4.8, transitivity and symmetry of "=" to derive equations.

$$\begin{aligned} (\mathtt{x} \wedge \mathtt{y}) \vee (\mathtt{x} \wedge \mathtt{z}) &= (\mathtt{x} \vee (\mathtt{x} \wedge \mathtt{z})) \wedge (\mathtt{y} \vee (\mathtt{x} \wedge \mathtt{z})) \ (\because L5) \\ &= ((\mathtt{x} \vee \mathtt{x}) \wedge (\mathtt{x} \vee \mathtt{z})) \wedge ((\mathtt{x} \vee \mathtt{y}) \wedge (\mathtt{y} \vee \mathtt{z})) \ (\because L5) \\ &= (\mathtt{x} \wedge (\mathtt{x} \vee \mathtt{z})) \wedge ((\mathtt{x} \vee \mathtt{y}) \wedge (\mathtt{y} \vee \mathtt{z})) \ (\because L1) \\ &= \mathtt{x} \wedge ((\mathtt{x} \vee \mathtt{y}) \wedge (\mathtt{y} \vee \mathtt{z})) \ (\because L4) \\ &= ((\mathtt{x} \wedge (\mathtt{x} \vee \mathtt{y})) \wedge (\mathtt{y} \vee \mathtt{z}) \ (\because L3) \\ &= \mathtt{x} \wedge (\mathtt{y} \vee \mathtt{z}) \ (\because \mathtt{L4}).\blacksquare \end{aligned}$$

As is well-known, we only need $L5$ as distributivity.

A *Boolean algebra* is an algebra $\langle B; F_{Bool}\rangle$, $F_{Bool} = \{\vee, \wedge, \_^{-1}, \mathtt{0}, \mathtt{1}\}$ such that $\mathtt{t}(\vee) = \mathtt{t}(\wedge) = 2$, $\mathtt{t}(\_^{-1}) = 1, \mathtt{t}(\mathtt{0}) = \mathtt{t}(\mathtt{1}) = 0$, $\langle B; \{\vee, \wedge\}\rangle$ is a distributive lattice and

$$\begin{array}{lll} Bo1: & \mathtt{0} \vee \mathtt{x} = \mathtt{x}, & \mathtt{1} \vee \mathtt{x} = \mathtt{x}, \\ Bo2: & \mathtt{x} \vee \mathtt{x}^{-1} = \mathtt{1}, & \mathtt{x} \wedge \mathtt{x}^{-1} = \mathtt{0} \end{array}$$

are both valid in it. We also review the famous valid equations: De Morgan's law:

$$(\mathtt{x} \vee \mathtt{y})^{-1} = \mathtt{x}^{-1} \wedge \mathtt{y}^{-1}, \quad (\mathtt{x}^{-1})^{-1} = \mathtt{x},$$

which we frequently use.

FACT 4.17 *PA and BA have subtype of Boolean algebra.*

Proof Clear. Set the map $\phi_B : F_{PA} \to F_{Bool}$ as

$$\phi_B((\_,\_)) = \vee,\ \phi_B([\_]) = \_^{-1},\ \text{and}\ \phi_B(\varepsilon) = 0. \blacksquare$$

Proposition 4.18 *PA and BA are weak algebras of Boolean algebra.*

Proof For a PA: $\langle A; F_{PA}\rangle$ and a Boolean algebra: $\langle B; F_{Bool}\rangle$, as $\langle A; F_{PA}\rangle$ has the subtype of $\langle B; F_{Bool}\rangle$, so we have the map $\phi_B : F_{PA} \to F_{Bool}$ such that $\mathtt{t}_{F_{PA}} = \mathtt{t}_{F_{Bool}} \circ \phi_B$ (Fact 4.17). Let

$$\hat{\phi}_B : T_{F_{PA}} \to T_{F_{Bool}}$$

be the mapping induced by ϕ_B. Applying $\hat{\phi}_B$ to both sides of "=" of all axioms $P1 \dots P5$ of PA we obtain

$$\begin{aligned}
P1' &: & \mathtt{x} \vee \mathtt{y} &= \mathtt{y} \vee \mathtt{x},\\
P2' &: & (\mathtt{x} \vee \mathtt{y}) \vee \mathtt{z} &= \mathtt{x} \vee (\mathtt{y} \vee \mathtt{z}),\\
P3' &: & (\mathtt{x}^{-1} \vee \mathtt{x})^{-1} &= 0,\\
P4' &: & ((\mathtt{x} \vee \mathtt{z})^{-1} \vee (\mathtt{y} \vee \mathtt{z})^{-1})^{-1} &= (\mathtt{x}^{-1} \vee \mathtt{y}^{-1})^{-1} \vee \mathtt{z},\ \text{and}\\
P5' &: & 0 \vee \mathtt{x} &= \mathtt{x}.
\end{aligned}$$

Similarly from the axioms of BA we have

$$\begin{aligned}
B1' &: & \mathtt{x} \vee \mathtt{y} &= \mathtt{y} \vee \mathtt{x},\\
B2' &: & (\mathtt{x} \vee \mathtt{y}) \vee \mathtt{z} &= \mathtt{x} \vee (\mathtt{y} \vee \mathtt{z}),\\
B3' &: & (\mathtt{x}^{-1} \vee \mathtt{y})^{-1} \vee \mathtt{x} &= \mathtt{x},\\
B4' &: & ((\mathtt{x} \vee \mathtt{z})^{-1} \vee (\mathtt{y} \vee \mathtt{z})^{-1})^{-1} &= (\mathtt{x}^{-1} \vee \mathtt{y}^{-1})^{-1} \vee \mathtt{z},\ \text{and}\\
B5' &: & 0 \vee \mathtt{x} &= \mathtt{x}.
\end{aligned}$$

If $i \neq 3$, Pi' is equal to Bi', so let us see the validity of $P1$, $P2$, $P4$, $P5$. Then we check if $P3'$ and $B3'$ are valid or not. Actually, $P1'$, $P2'$ are trivially valid. As for $P4'$,

$$\begin{aligned}
((\mathtt{x} \vee \mathtt{z})^{-1} \vee (\mathtt{y} \vee \mathtt{z})^{-1})^{-1} &= (\mathtt{x} \vee \mathtt{z}) \wedge (\mathtt{y} \vee \mathtt{z})\ \ (\because \text{De Morgan's law})\\
&= (\mathtt{x} \wedge \mathtt{y}) \vee \mathtt{z}\ \ (\because \text{distributivity})\\
&= (\mathtt{x}^{-1} \vee \mathtt{y}^{-1})^{-1} \vee \mathtt{z}\ \ (\because \text{De Morgan's law})
\end{aligned}$$

$P5'$ also is trivial. For $P3'$, it follows from $\mathtt{x}^{-1} \vee \mathtt{x} = 1$ and $1^{-1} = 0$. Finally, $B3'$: $(\mathtt{x}^{-1} \vee \mathtt{y})^{-1} = (\mathtt{x} \wedge \mathtt{y}^{-1}) \vee \mathtt{x}$ by De Morgan's law, and $L4$ yields the result. ■

Given $\langle A; F\rangle$, a polynomial $\mathtt{p} \in T_F$ is said to have *occurrences of operator symbols* $\Psi \subseteq F$ if for each $\mathtt{f} \in \Psi$ it occurs in $\mathtt{p}$ in the process of building $\mathtt{p}$.

Definition 4.19 Let a couple of algebras $\langle A; F\rangle$ and $\langle A; G\rangle$ such that $F \subset G$, adjoined with axioms for each be given (let them be denoted by $\mathtt{A_F}$ *and* $\mathtt{A_G}$, respectively). For an operator symbol $\mathtt{g} \in G$, we say $\Phi_{\mathtt{g}} = \{\cdots, \mathtt{f_i}, \cdots\} \subseteq F$ *implements* $\mathtt{g}$ provided: There exists $\mathtt{p_g} \in T_F$ with occurrences $\Phi_{\mathtt{g}}$ that satisfies

$$\mathtt{p_g} = \mathtt{g} \text{ is valid in } \langle A; G\rangle, \text{ and}$$

$$\text{for any axiom } \mathtt{a_g} \in \mathtt{A_G}, \ \ \mathtt{a_g(p_g/g)} \text{ is valid equation in } \langle A; F\rangle$$

If there is $\Phi_{\mathtt{g}} \subseteq F$ that implements $\mathtt{g}$ for arbitrary $\mathtt{g} \in G$, we say the algebra $\langle A; F\rangle$ implements $\langle A; G\rangle$. We also express "$\mathtt{q_g}$ implements $\mathtt{g}$" if the occurring Φ of $\mathtt{q_g} \in T_F$ implements $\mathtt{g}$. The notation $[\mathtt{p}]_{\mathtt{g}}$ is used to denote all polynomials in T_F that implements $\mathtt{g}$.

Later on we will see the fact by observing derivations in the "rewritten" version of PA and BA.

Suppose $\langle A; F\rangle$ implements $\langle A; G\rangle$. For $\mathtt{p} \in T_G$ there exists $G|_{\mathtt{p}} = \{\cdots, \mathtt{g_i}, \cdots\}_{i\in I} \subseteq F$, which consists of operator symbols of G occurring in $\mathtt{p}$, adjoined with $[\mathtt{p}]_{\mathtt{g_i}}$ for each $i \in I$. Let us adopt the notation $\mathtt{p_i^F}$ to denote elements of $[\mathtt{p}]_{\mathtt{g_i}}$ and suppose $\mathtt{p_i^F}$ consists of $\{\cdots, \mathtt{f_j^i}, \cdots\}_{j\in J}$. Then we have a polynomial $\mathtt{p}^F \in T_F$, which consists of

$$\bigcup_{i\in I}\{\cdots, \mathtt{f_j^i}, \cdots\}_{j\in J}$$

and

$$\overline{Denot}_F(\mathtt{p}^F) = \overline{Denot}_G(\mathtt{p}).$$

Note that if the arity of $\mathtt{p} \in T_F$ is equal to that of $\mathtt{q} \in T_F$, then

$$\overline{Denot}_F(\mathtt{p}) = \overline{Denot}_F(\mathtt{q}) \text{ if and only if } \mathtt{p} = \mathtt{q} \text{ is valid },$$

therefore we state:

Lemma 4.20 *Let T_G^{γ} be a representation of the set of γ-ary polynomials in $\langle A; G\rangle$. If $\langle A; G\rangle$ is implemented by $\langle A; F\rangle$, then for all $\mathtt{p}, \mathtt{q} \in T_G^{\gamma}$:*

$$\textit{If } \mathtt{p} = \mathtt{q} \textit{ is valid in } \langle A; G\rangle, \textit{ then } \mathtt{p^F} = \mathtt{q^F} \textit{ is valid in } \langle A; F\rangle.$$

Proof $\mathtt{p} = \mathtt{q}$ is valid in $\langle A; G\rangle$ iff $\overline{Denot}_G(\mathtt{p}) = \overline{Denot}_G(\mathtt{q})$. On the other hand, $\overline{Denot}_F(\mathtt{p}^F) = \overline{Denot}_G(\mathtt{p})$, $\overline{Denot}_F(\mathtt{q}^F) = \overline{Denot}_G(\mathtt{q})$ hold, respectively. ∎

6.4.3 PA and Boolean Algebra

Let us introduce a "change on notation" just for convenience. The *rewritten PA* is the algebra
$\langle A; \{\phi_B((\_,\_)), \phi_B([\_]), \phi_B(\varepsilon)\}\rangle$ where ϕ_B is the mapping defined in Fact 4.17. Namely, we call an algebra $\langle A;\ F\rangle$ a rewritten PA, denoted by $\widehat{PA}$ if

$$F = \{\phi_B((\_,\_)), \phi_B([\_]), \phi_B(\varepsilon)\} = \{\vee, \_^{-1}, 0\}$$

adjoined with the axioms (as were seen already):

$$\begin{aligned}
P1' &: & \mathtt{x} \vee \mathtt{y} &= \mathtt{y} \vee \mathtt{x},\\
P2' &: & (\mathtt{x} \vee \mathtt{y}) \vee \mathtt{z} &= \mathtt{x} \vee (\mathtt{y} \vee \mathtt{z}),\\
P3' &: & (\mathtt{x}^{-1} \vee \mathtt{x})^{-1} &= \mathtt{0},\\
P4' &: & ((\mathtt{x} \vee \mathtt{z})^{-1} \vee (\mathtt{y} \vee \mathtt{z})^{-1})^{-1} &= (\mathtt{x}^{-1} \vee \mathtt{y}^{-1})^{-1} \vee \mathtt{z}, \text{ and}\\
P5' &: & \mathtt{0} \vee \mathtt{x} &= \mathtt{x}.
\end{aligned}$$

Theorem 4.21 *For a Boolean algebra* $\mathcal{B} = \langle B; \{\vee, \wedge, \_^{-1}, 0, 1\}\rangle$*, the weak algebra of* $\mathcal{B}$: $\langle B; \{\vee, \_^{-1}, 0\}\rangle$ *is* $\widehat{PA}$.

Proof Due to Proposition 4.18. ■

Proposition 4.22 *In a* $\widehat{PA}$*, all of the following equations are valid:*

$$\begin{aligned}
&(a) & \mathtt{x} \vee \mathtt{x} &= \mathtt{x}, & (\mathtt{x}^{-1} \vee \mathtt{x}^{-1})^{-1} &= \mathtt{x},\\
&(b) & (\mathtt{x}^{-1} \vee (\mathtt{x} \vee \mathtt{y})^{-1})^{-1} &= \mathtt{x}, & \mathtt{x} \vee (\mathtt{x}^{-1} \vee \mathtt{y}^{-1})^{-1} &= \mathtt{x},\\
&(c) & \mathtt{x} \vee (\mathtt{y}^{-1} \vee \mathtt{z}^{-1})^{-1} &= ((\mathtt{x} \vee \mathtt{y})^{-1} \vee (\mathtt{x} \vee \mathtt{z})^{-1}), & &\\
&(d) & \mathtt{x} \vee \mathtt{x}^{-1} &= \mathtt{0}^{-1}, & (\mathtt{x}^{-1} \vee (\mathtt{x}^{-1})^{-1})^{-1} &= \mathtt{0}^{-1},\\
&(e) & (\mathtt{x}^{-1})^{-1} &= \mathtt{x}. & &
\end{aligned}$$

Proof Let us observe the demonstration of (e). (Others are basically straightforward.)

$$\begin{aligned}
(\mathtt{x}^{-1})^{-1} &= \mathtt{0} \vee (\mathtt{x}^{-1})^{-1}\\
&= ((\mathtt{x}^{-1})^{-1} \vee \mathtt{x}^{-1})^{-1} \vee (\mathtt{x}^{-1})^{-1}\\
&= ((\mathtt{x}^{-1} \vee (\mathtt{x}^{-1})^{-1})^{-1} \vee (\mathtt{x} \vee (\mathtt{x}^{-1})^{-1})^{-1})^{-1}\\
&= (\mathtt{0} \vee (\mathtt{x} \vee (\mathtt{x}^{-1})^{-1})^{-1})^{-1}\\
&= ((\mathtt{x}^{-1} \vee \mathtt{x})^{-1} \vee (\mathtt{x} \vee (\mathtt{x}^{-1})^{-1})^{-1})^{-1}\\
&= ((\mathtt{x}^{-1})^{-1} \vee ((\mathtt{x}^{-1})^{-1})^{-1})^{-1} \vee \mathtt{x}\\
&= \mathtt{0} \vee \mathtt{x}\\
&= \mathtt{x}. \ ■
\end{aligned}$$

Note that all of the equations are those that are familiar "theorems" held in ordinary Boolean algebra.

Let us not go further for $\widehat{PA}$, whereas $\widehat{BA}$ would be a good material to show the technical detail to indicate the standpoint of the context.

6.4.4 BA and Distributive Lattice

The notation $\widehat{BA}$ is also used to denote rewritten Brownian algebra. Let us review the axioms of $\widehat{BA}$:

$$\begin{aligned}
B1' &: & \mathtt{x} \vee \mathtt{y} &= \mathtt{y} \vee \mathtt{x}, \\
B2' &: & (\mathtt{x} \vee \mathtt{y}) \vee \mathtt{z} &= \mathtt{x} \vee (\mathtt{y} \vee \mathtt{z}), \\
B3' &: & (\mathtt{x}^{-1} \vee \mathtt{y})^{-1} \vee \mathtt{x} &= \mathtt{x}, \\
B4' &: & ((\mathtt{x} \vee \mathtt{z})^{-1} \vee (\mathtt{y} \vee \mathtt{z})^{-1})^{-1} &= (\mathtt{x}^{-1} \vee \mathtt{y}^{-1})^{-1} \vee \mathtt{z}, \text{ and} \\
B5' &: & \mathtt{0} \vee \mathtt{x} &= \mathtt{x}.
\end{aligned}$$

Proposition 4.23 *In* $\widehat{BA}$,

$$(\mathtt{x}^{-1})^{-1} = \mathtt{x} \textit{ is valid.}$$

Proof

$$\mathtt{x}^{-1} = ((\mathtt{x}^{-1})^{-1} \vee \mathtt{x})^{-1} \vee \mathtt{x}^{-1} \quad (\because B3')$$

$$\downarrow$$

$$\begin{aligned}
(\mathtt{x}^{-1})^{-1} &= (((\mathtt{x}^{-1})^{-1} \vee \mathtt{x})^{-1} \vee \mathtt{x}^{-1})^{-1} \\
&= (((\mathtt{x}^{-1})^{-1} \vee \mathtt{x})^{-1} \vee ((\mathtt{x}^{-1})^{-1} \vee \mathtt{x})^{-1})^{-1} \quad (\because B3') \\
&= (\mathtt{x}^{-1} \vee \mathtt{x}^{-1})^{-1} \vee (\mathtt{x}^{-1})^{-1} \quad (\because B4') \\
&= ((\mathtt{0} \vee \mathtt{x})^{-1} \vee (\mathtt{0} \vee \mathtt{x})^{-1})^{-1} \vee (\mathtt{x}^{-1})^{-1} \quad (\because B5') \\
&= ((\mathtt{0}^{-1} \vee \mathtt{0}^{-1})^{-1} \vee \mathtt{x}) \vee (\mathtt{x}^{-1})^{-1} \quad (\because B4') \\
&= (((\mathtt{0}^{-1} \vee \mathtt{0}^{-1})^{-1} \vee \mathtt{0}) \vee \mathtt{x}) \vee (\mathtt{x}^{-1})^{-1} \quad (\because B5') \\
&= (\mathtt{0} \vee \mathtt{x}) \vee (\mathtt{x}^{-1})^{-1} \quad (\because B3') \\
&= \mathtt{x} \vee (\mathtt{x}^{-1})^{-1} \quad (\because B5') \\
&= \mathtt{x} \vee (\mathtt{x}^{-1} \vee \mathtt{0})^{-1} \quad (\because B5') \\
&= \mathtt{x} \quad (\because B3'). \blacksquare
\end{aligned}$$

Proposition 4.24 *In* $\widehat{BA}$,

$$\mathtt{x} \vee \mathtt{x} = \mathtt{x} \quad \textit{is valid.}$$

Proof

$$\begin{aligned} \mathtt{x} \vee \mathtt{x} &= (\mathtt{x}^{-1})^{-1} \vee \mathtt{x} \\ &= (\mathtt{x}^{-1} \vee 0)^{-1} \vee \mathtt{x} \\ &= \mathtt{x}. \blacksquare \end{aligned}$$

Proposition 4.25 *In* $\widehat{BA}$,

$$(a)\ \mathtt{x} \vee (\mathtt{x}^{-1} \vee \mathtt{y}^{-1})^{-1} = \mathtt{x} \textit{ and}$$

$$(b)\ (\mathtt{x}^{-1} \vee (\mathtt{x} \vee \mathtt{y})^{-1})^{-1} = \mathtt{x} \quad \textit{are valid.}$$

Proof

$$\begin{aligned} (a):\ \mathtt{x} \vee (\mathtt{x}^{-1} \vee \mathtt{y}^{-1})^{-1} &= ((\mathtt{x} \vee \mathtt{x})^{-1} \vee (\mathtt{x} \vee \mathtt{y})^{-1})^{-1} \\ &= (\mathtt{x}^{-1} \vee (\mathtt{x} \vee \mathtt{y})^{-1})^{-1} \\ &= (\mathtt{x}^{-1} \vee ((\mathtt{x}^{-1})^{-1} \vee \mathtt{y})^{-1})^{-1} \\ &= (\mathtt{x}^{-1})^{-1} \\ &= \mathtt{x}. \end{aligned}$$

$$\begin{aligned} (b):\ &(a) \cdots \mathtt{x} \vee (\mathtt{x}^{-1} \vee \mathtt{y}^{-1})^{-1} = \mathtt{x} \\ &\downarrow \\ &\mathtt{x}^{-1} \vee ((\mathtt{x}^{-1})^{-1} \vee (\mathtt{y}^{-1})^{-1})^{-1} = \mathtt{x}^{-1} \\ &\downarrow \\ &\mathtt{x}^{-1} \vee (\mathtt{x} \vee \mathtt{y})^{-1} = \mathtt{x}^{-1} \\ &\downarrow \\ &(\mathtt{x}^{-1} \vee (\mathtt{x} \vee \mathtt{y})^{-1})^{-1} = (\mathtt{x}^{-1})^{-1} = \mathtt{x},\ \blacksquare \end{aligned}$$

Proposition 4.26 *In* $\widehat{BA}$,

$$(\mathtt{z}^{-1} \vee (\mathtt{x} \vee \mathtt{y})^{-1})^{-1} = (\mathtt{x}^{-1} \vee \mathtt{z}^{-1})^{-1} \vee (\mathtt{y}^{-1} \vee \mathtt{z}^{-1})^{-1} \quad \textit{is valid.}$$

Proof

$$
\begin{aligned}
((\mathtt{x}^{-1} \vee \mathtt{z}^{-1})^{-1} \vee (\mathtt{y}^{-1} \vee \mathtt{z}^{-1})^{-1})^{-1} &= ((\mathtt{x}^{-1})^{-1} \vee (\mathtt{y}^{-1})^{-1})^{-1} \vee \mathtt{z}^{-1} \\
&= (\mathtt{x} \vee \mathtt{y})^{-1} \vee \mathtt{z}^{-1} \\
&\downarrow \\
(((\mathtt{x}^{-1} \vee \mathtt{z}^{-1})^{-1} \vee (\mathtt{y}^{-1} \vee \mathtt{z}^{-1})^{-1})^{-1})^{-1} &= (\mathtt{x}^{-1} \vee \mathtt{z}^{-1})^{-1} \vee (\mathtt{y}^{-1} \vee \mathtt{z}^{-1})^{-1} \\
&= ((\mathtt{x} \vee \mathtt{y})^{-1} \vee \mathtt{z}^{-1})^{-1}. \blacksquare
\end{aligned}
$$

Theorem 4.27 *A $\widehat{BA}$ always implements a distributive lattice, that is, let $\widehat{BA} = \langle A; \{\vee', \_^{-1}, 0\}\rangle$ and let us introduce a new operator symbol "$\wedge'$" such that*

$$\overline{Denot}(\mathtt{x} \wedge' \mathtt{y}) = \overline{Denot}((x^{-1} \vee' y^{-1})^{-1}).$$

Then, the algebra $\langle A; \{\vee', \wedge'\}\rangle$ is a distributive lattice.

Proof We have already observed Propositions 4.24, 4.25, and 4.26, which ensured that $L1 \ldots L5$ are all valid in the algebra $\langle A; \{\vee', \wedge'\}\rangle$. ■

The algebra $\widehat{BA}$ is nothing other than a special case of distributive lattice. "Special case" means that it has nullary operator "0" with one of the axioms

$$\mathtt{x} \vee \mathtt{0} = \mathtt{x}.$$

Interestingly, it looks that the axioms of $\widehat{PA}$ do not derive $B3'$:

$$(\mathtt{x}^{-1} \vee \mathtt{y})^{-1} \vee \mathtt{x} = \mathtt{x},$$

and vice versa: Axioms $P3'$ might be independent from $\widehat{BA}$:

$$(\mathtt{x}^{-1} \vee \mathtt{x})^{-1} = \mathtt{0}.$$

These independencies are still left not given clear proof. (The original book (Spencer Brown, 1969) did not present clear statements for the issue.)

6.5 Conclusion: Implications

As the author mentioned in the introductory section, we have seen the formulations of our subject *G.Spencer Brown's theme* on two lines. The first line was a logical approach.

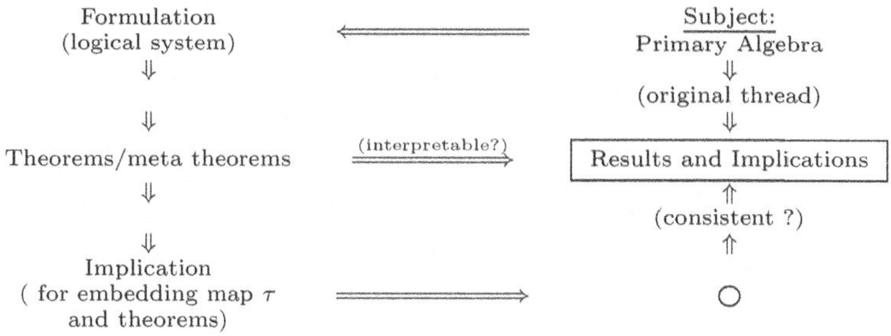

Fig. 6.2 First story

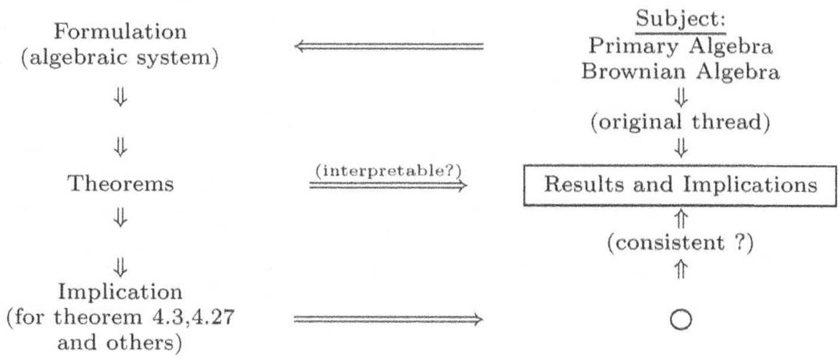

Fig. 6.3 Second story

The implications seem to depend mostly on the mapping τ (and its inverse σ). It was very natural and even trivial to define them, and we found that it is a sort of embedding injection into the wffs of propositional calculus.

In the second line, we introduced universal algebraic context to formulate our subject (Primary Algebra and Brownian Algebra.)

Apart from primary algebra, the so-called *Brownian algebra* has not attracted researchers' attention. Theorem 4.27 may be one resource to understand what it precisely is.

So, how could we explain the interpretability and consistency with original results and implication ? As has been observed since 1990s, researchers (especially in the area of social science) have been keen on finding out "new" framework and weapon to grapple with big and complex systems. It seemed that the struggle was the same thing as was emerged in the battle between holism and reductionism. Succeedingly we observed the theory of hierarchy and complex systems, both of which may have been sort of arbitration. [3]

Under the circumstance, G.Spencer Brown presented the notion of "calculus of indication" which, at first, was thought of as an original and essential way to deal with the foundation of "cognition." But as we already saw in this paper, they can be re-formulated into familiar mathematical entities, by practicing as Figs. 6.2 and 6.3.

[3] Actually, G.Spencer Brown and F.J.Varela wrote in quite a holismic viewpoint.

The impact created by the movement caused a particular kind of epistemological standpoint's spread. The standpoint was, namely, the hybrid view of "structuralism" and "post-structuralism." Then the so-called postmodernism movement emerged, and wide acceptance of it occurred. As is well-known, the method of "postmodernism" in making scientific statements was severely criticized by Alan Sokal, and after that, the movement has been under severe criticism.

The author would like to leave semantical/philosophical epistemology, for the discussion is another thing to be done separately in more careful way.

References

Graetzer, G. (1979). *Universal algebra* (2nd ed.). Springer.

Spencer Brown, G. (1969). *Laws of form*. George Allen and Unwin.

Schwartz, D. G. (1979). Isomorphisms of spencer Brown's laws of form and Varela's calculus for self reference. *International Journal of General Systems*, *6*, 239–255.

Shimogawa, T. (1998). The mathematical foundation of 'calculus of indication'. *Musashi Daigaku Ronshuu*, *46*(1), 69–101.

Shimogawa, T., & Takara, Y. (1994). Reconstruction of g.spencer brown's theme. *International Journal of General Systems*, *23*(1), 2–21.

Varela, F. J. (1979). *Principle of biological autonomy*. Elsevier North Holland.

Part II
Systems Modeling

Chapter 7
Mutual Learning Process Model in Soft Game Perspective

Kyoichi Kijima

Abstract Mutual learning process is dynamic process of transferring information and internalizing this information by relevant players often with essentially diversified and conflicting preferences on the outcomes. This article identifies four stages in mutual learning process depending on the degree of mutual understanding of two players and analyzes transitions among them.

The purpose of the present article is twofold: The first is to analyze and formulate dynamic mutual learning process of two players in soft game theoretic perspective. In order to overcome limitations of the analytical and engineering approach so far as well as to describe a soft problematic situation more realistically, we adopt ideas of hypergame analysis as a basis of our investigation to deal with subjective perception of the players. The developed model, which is called Mutual Learning Process Model (MLPM), not only assumes that the players possess essentially diversified preferences and perceive even the same problematic situation differently and subjectively, but also focuses on stage transitions of the learning process. Each stage is featured by the internal model and concept of rationality adopted by the players. We argue how perception of and attitude toward the situation changes dynamically and how "emergent properties" arise, as the learning process advances.

The second is to apply MLPM to an illustrative example of the Gulf War from 1990 to 1991. The application provides practical insights about the actual problematic situation as well as demonstrates theoretical capability of the model.

Keywords Soft problematic situation · Mutual learning process · Hypergames · Hyper-Nash equilibrium · Internal model · What-if analysis

---

Some basic ideas of this article were developed in K. Kijima (1996).

K. Kijima (✉)
School of Business Management, Bandung Institute of Technology, Bandung, Indonesia
e-mail: kijima@kijima-lab.com

K. Kijima et al. (eds.), *Systems Research I*, Translational Systems Sciences 26,
https://doi.org/10.1007/978-981-16-9937-5_7

7.1 Introduction

If a decision situation involves a variety of people with essentially diversified and conflicting preferences on the outcomes, it is often referred to as a soft problematic situation in the fields of systems science and Operational Research (Checkland, 1981; Rosenhead, 1989). This article focuses on mutual learning process in such a soft problematic situation and analyzes it through constructing a rigorous model of it.

Mutual learning is process of transferring information and internalizing this information by relevant players often with essentially diversified and conflicting preferences on the outcomes. Since in the process misunderstanding, misperception, and/or complete mutual understanding of the players could happen, we may identify several stages in it and transitions among them.

Game theory is one of analytical frameworks that can accommodate such problematic situations. The theory usually analyzes the rational behavior of players under the assumption that the players can observe the problematic situation objectively but subjective observation by the players is basically out of the research scope. It, however, does not necessarily seem to reflect reality, so that we try to relax this limitation in our modeling.

Techniques such as reasoning under uncertainty, fuzzy sets, decision trees, and blackboard architectures have also been developed for multi-player decision making in the field of artificial intelligence (Neches et al., 1991). The main concerns in these areas, however, do not necessarily seem to be with subjective perception and/or misperception by the players, though such perception problems are crucial in investigating mutual learning.

This article assumes in mutual learning process, rationality, rather than emotion and feeling, plays a crucial role for describing players' behavior. The players are presumed to behave rationally based on their subjective internal model at each stage, to rewrite it dynamically and to change their concept of "rationality" by learning, as time goes on. Internal model (which is sometimes interchangeably called as mental model or interpretive model) means a representation of reality within the mind of an individual (Wonham, 1976).

The purpose of the present article is twofold: The first is to analyze and formulate dynamic learning process of two players in soft game theoretic perspective (Bryant, 2015). In order to overcome the limitations of the analytical and engineering approach as well as to describe soft problematic situations more realistically, we adopt ideas of hypergame analysis as a basis of our investigation to deal with subjective perception of the players (Kovach et al., 2015).

The fundamental idea of hypergame analysis is that the players may conceptualize problems in a similar manner to that of game theory, but they see different games. That is, it formulates interactive decision situations not by a single game, but by a collection of subjective games of each player. We model a dynamic mutual learning process by extending hypergame analysis in a unified way to propose a simple and intuitive model, i.e., MLPM (Mutual Learning Process Model), for examining it.

The second is to apply MLPM to an illustrative example of Gulf War from 1990 to 1991 not only to provide realistic insights about the practical problematic situation, but also to demonstrate theoretical capability of the model.

The structure of this article is as follows: In Sect. 7.2 we will analyze mutual learning process of two players to identify four stages in it, i.e., isolated, symbiotic, value sharing, and complete mutual understanding stages as well as transition path among them. In Sect. 7.3 we propose MLPM, a formal model of mutual learning process, based on the analysis in the previous section. In Sect. 7.4 we apply MLPM to an illustrative example of the Gulf War 1990–1991. Finally, Sect. 7.5 gives concluding remarks and suggests further research questions.

7.2 Four Stages of Mutual Learning

This article focuses on mutual learning process between two players, where we assume the players involved in a soft problematic situation. MLPM presumes that in a soft problematic situation it is almost impossible for the players to solve problems by collecting all the relevant information to select a solution on the basis of optimality. Rather, improvement of the situation and increase of knowledge by learning is essential and is probably the only way to tackle such a situation. Along mutual learning process the players change their understanding (internal model) of the situation by mutual interaction with each other.

Basic assumptions on the players are as follows:

1. Both the players are symmetric: Both the players follow the same behavioral pattern and logic.
2. Each player subjectively constructs his/her internal model relevant to the problematic situation around him/her.
3. Each player makes decisions for attaining his/her goal by referring to his/her internal model based on his/her "rationality."
4. Each player dynamically rewrites its internal model along the time horizon and changes its concept of rationality by learning through mutual interaction.

We assume mutual learning process consists of four stages, namely, isolated stage, mutual interpretation stage, value sharing stage, and completed mutual understanding stage as well as transition paths among them.

Let us denote the players p and q and assume they both are involved in some soft problematic situation.

7.2.1 Isolated Stage

At the present stage p and q have no particular knowledge about each other and they construct their own "subjective" internal model completely independently based on

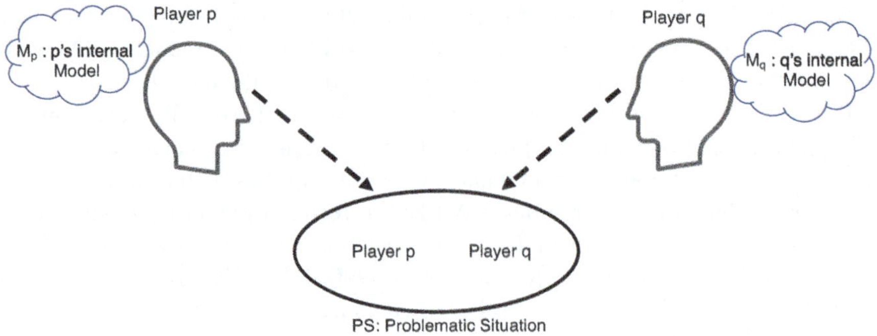

Fig. 7.1 Isolated stage

their observation about the situation. A typical example of this stage may occur when two players encounter at the first time without any knowledge about the other.

We assume p to perceive the problematic situation to construct her internal model M_p subjectively, while q is symmetrically assumed to perceive his environment to construct his internal model M_q subjectively. The constructed models, M_p and M_q, are generally independent of each other, and it is quite common for the players to perceive even the same phenomenon in *PS* in a different way (Refer to Fig. 7.1). Since their internal models are constructed independently in such an isolated circumstance, they might be completely different and not correlated. They examine their subjective model by adopting their rationality concept and may make decisions to take actions at this stage.

7.2.2 *Mutual Interpretation Stage*

At this stage we assume not only that players p and q have constructed subjectively their internal model M_p and M_q, respectively, but also that there exists some sort of interpretation between the two players. That is, even though they use different language, they understand what they refer to by different terms (Refer to Fig. 7.2).

For example, suppose p says an "apple" and q says "ringo" meaning an apple in Japanese. If p understands ringo means an apple in her language, then we can say she interprets correctly. We should notice that interpretation is not always correct, and there are right and wrong interpretations.

They examine their subjective model by adopting their rationality concept and may make decisions to take actions at this stage.

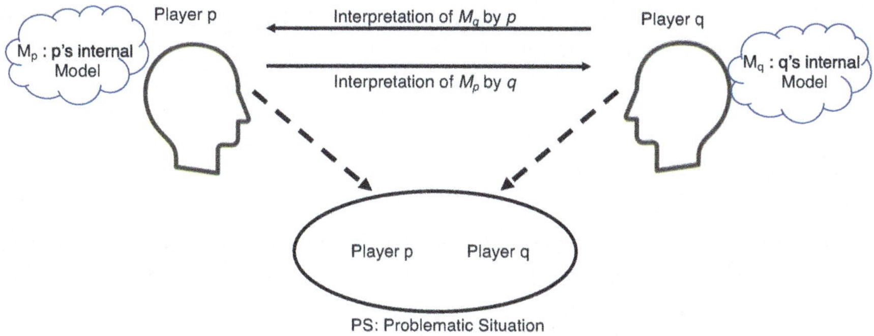

Fig. 7.2 Mutual interpretation stage

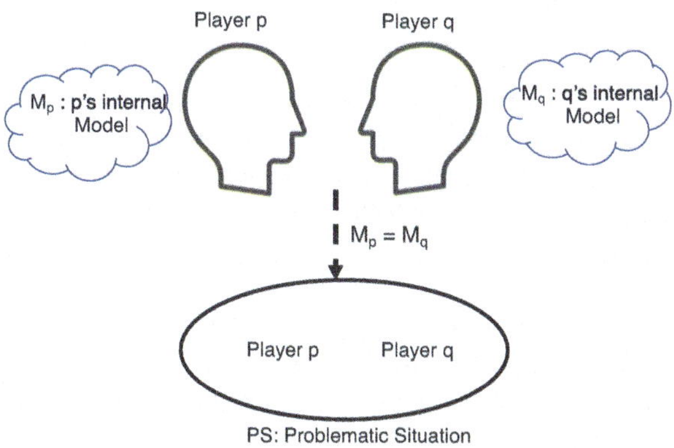

Fig. 7.3 Complete mutual understanding stage

7.2.3 Value Sharing Stage

When both the players understand correctly the counterpart's set of strategies and preference ordering, we say they share the same value system and call the situation value sharing stage. At this stage they may use their own language, but both the players have correct interpretation of each other (Fig. 7.3).

They examine their subjective model by adopting their rationality concept and may make decisions to take actions at this stage.

7.2.4 Complete Mutual Understanding Stage

This is an ultimate stage of value sharing where they understand the counterpart's strategies and preference ordering completely. In other words, at this stage they share the same internal model as common knowledge.

They examine their subjective model by adopting their rationality concept and may make decisions to take actions at this stage.

7.2.5 State Transitions

The four stages in mutual learning process are characterized in terms of degree and quality of the mutual relationship between the players; i.e., none, some, consistent, and complete. Though, in order to examine mutual learning process, we also need to focus on how the stages transit from one to another. One of our interests concerning the transition pass is to investigate under what condition there might be emergence of new solutions or, oppositely, disappearance of existing solutions along the path (Fig. 7.4).

The transition is caused by mutual interaction when they increase/decrease knowledge about the situation. When they do not make any decision nor take any action, both the players may rewrite their internal model and change their concept of rationality by mutual interaction to move to another stage, as time goes by.

The process does not necessarily follow the four stages step-by-step in this order along the outer cycle in Fig. 7.4. The arrows of Fig. 7.4 illustrate a stage may move to any other stage as time goes by.

Though, it may be natural to assume the process begins with Isolated Stage, where the players perceive a problematic situation involving them in an independent and isolated way. They examine their subjective internal model by adopting their

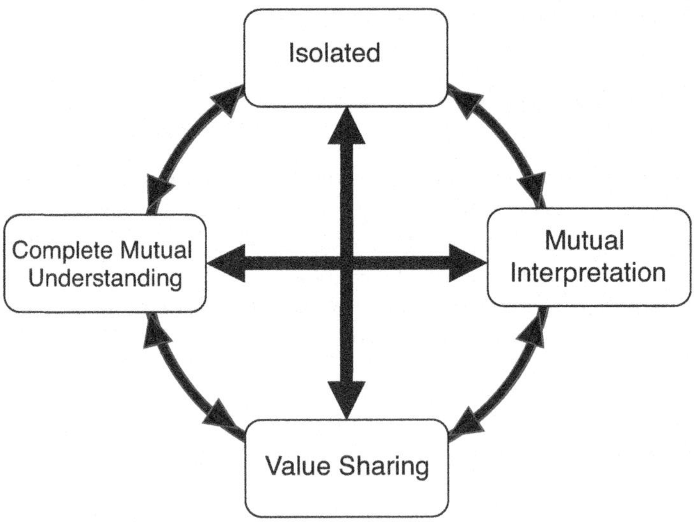

Fig. 7.4 Four stages and state transitions

rationality concept. They may make decisions to take actions at this stage, or transit to another stage such as mutual interpretation stage.

7.3 MLPM: Mutual Learning Process Model

In this section we formally develop Mutual Learning Process Model (MLPM) representing a dynamic learning process between two players who perceive and interpret the problematic situation subjectively in soft game theoretic perspective. It provides an intuitive framework for understanding actual soft problematic situations.

MLPM is represented by a sequence of decision situations, each of which is characterized by the internal model and rationality concept associated with each player as well as by the overall rationality. We are interested in what changes happen along the sequence.

Fundamental assumptions of MLPM are formulated as follows:

First, player i, where $i = p$ or q, is assumed to be associated with an internal model M_i and a decision criterion or rule R_i. M_i is formulated by a strategic game form representing i's subjective understanding of the environmental surrounding. M_i is usually constructed through and changes by interactions with the environment including the other player.

The decision criterion R_i tells player i what alternatives are reasonable or rational in the perceived environment. R_i represents i's concept of rationality in this sense. The decision making situation of player i is then characterized by the pair (M_i, R_i).

Let us denote "overall rationality" by X, which determines the outcomes of the decision situation as a whole. Then the overall decision situation can be described by

$$((M_p, R_p), (M_q, R_q), X).$$

Secondly, MLPM explicitly deals with dynamic transition of perception of the decision situation by the players. Indeed, an overall decision situation $((M_p, R_p), (M_q, R_q), X)$ may change to another $((M'_p, R'_p), (M'_q, R'_q), X')$ as a result of dynamic interaction between the players, as time goes on. This change may continue forever at least in principle.

To formulate such dynamic changes, MLPM adopts several models such as simple hypergame and noncooperative game in a unified way.

As discussed previously, we assume mutual learning process consists of four stages in it, i.e., Isolated, Mutual Interpretation, Value Sharing, and Complete Mutual Understanding. At isolated stage, hypergame analysis should be useful and insightful, but at complete mutual understanding stage, a conventional noncooperative game would provide sharper results. In MLPM, indeed, these four stages are represented by simple hypergame, symbiotic hypergame, hypergame with the same value systems, and noncooperative game, respectively (Refer to Fig. 7.5).

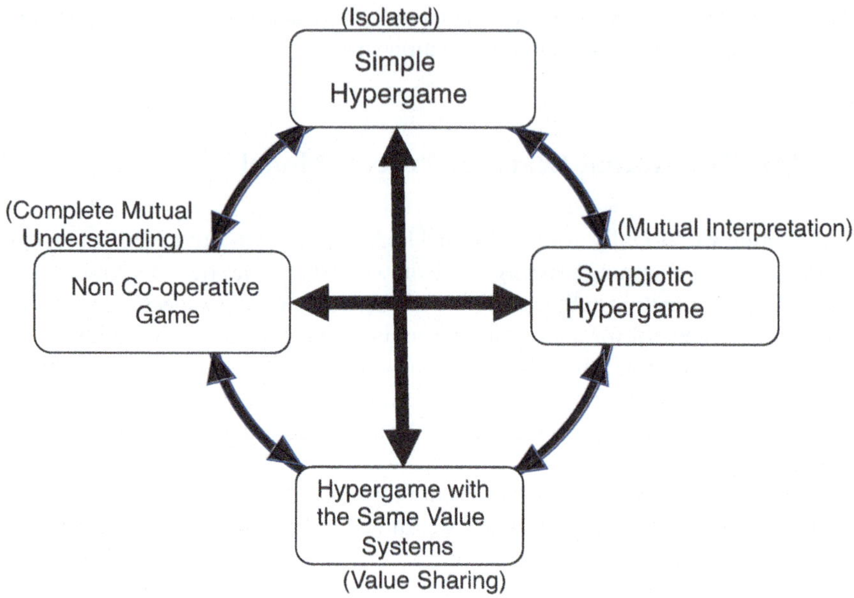

Fig. 7.5 Mutual learning process model (MLPM)

Each stage is characterized by the internal models and rationality concept associated with the players.

MLPM considers the noncooperative game situation as the ultimate stage where mutual understanding and learning of the players has been accomplished. However, since it works as a basis for formulating other stages, we will begin with the well-known two person noncooperative game.

7.3.1 Noncooperative Game

Definition 3.1 A two person noncooperative game with players p and q is a quadruple $G = (S_p, S_q, \succeq_p, \succeq_q)$.

For $i \in \{p, q\}$, S_i denotes a set of strategies of player i, while $\succeq_i$ is i's preference ordering on $S_p \times S_q$. For any (s_p, s_q) and $(s'_p, s'_q) \in S_p \times S_q$, $(s_p, s_q) \succeq_i (s'_p, s'_q)$ means that i prefers (s_p, s_q) to (s'_p, s'_q) or is indifferent between (s_p, s_q) and (s'_p, s'_q). We assume that S_p and S_q are finite while $\succeq_p$ and $\succeq_q$ are linear orderings represented by some ordinal utility functions.

We adopt the well-known Nash equilibrium for describing rationality for a two person noncooperative game (for example, see Gibbons, 1992).

Let us denote the class of two person noncooperative games that we are interested in by

$$\mathcal{G} = \{G = (S_p, S_q, \succeq_p, \succeq_q) | S_p \subset \mathcal{S}_p, S_q \subset \mathcal{S}_q,$$

$$\succeq_p \subset (S_p \times S_q) \times (S_p \times S_q), \succeq_q \subset (S_p \times S_q) \times (S_p \times S_q)\}$$

where $\mathcal{S}_p$ and $\mathcal{S}_q$ are given supersets of S_p and S_q, respectively.

Definition 3.2 Nash criterion is a function N from $\mathcal{G}$ into $\mathcal{P}(S_p \times S_q)$ such that for $G = (S_p, S_q, \succeq_p, \succeq_q)$ in $\mathcal{G}$, $s^* = (s_p^*, s_q^*)$ is in $N(G)$ if and only if we have

1. $s^* = (s_p^*, s_q^*)$ is in $S_p \times S_q$.
2. $(\forall s_p \in S_p)((s_p^*, s_q^*) \succeq_p (s_p, s_q^*))$ and $(\forall s_q \in S_q)((s_p^*, s_q^*) \succeq_q (s_p^*, s_q))$ hold.

$\mathcal{P}(S_p \times S_q)$ denotes the powerset of $S_p \times S_q$.

An element in $N(G)$ is called a Nash equilibrium of G. The definition implies that if $s^* = (s_p^*, s_q^*) \in S_p \times S_q$ is a Nash equilibrium, then there is no incentive for either of the players to change their strategy as long as the other does not change its strategy.

At this stage both the players are associated with (G, N), where they completely share the same model of the problematic situation, though they compete with each other. Noncooperative game situation is represented by $((G, N), (G, N); \phi)$, where ϕ indicates that overall rationality X does not exist explicitly.

7.3.2 *Simple Hypergame*

Simple hypergame (Bennett, 1980; Bennett and Dando, 1979; Bennett et al., 1980, 1989) assumes both the players to perceive the problematic situation independently and they have no interactions between them.

Definition 3.3 A simple hypergame of players p and q is a pair of (G_p, G_q), where $G_p = (S_p, S_{qp}, \succeq_p, \succeq_{qp})$ is a game that p believes both sides perceive, while $G_q = (S_{pq}, S_q, \succeq_{pq}, \succeq_q)$ is a game that q believes both sides perceive.

In G_p, S_p denotes a set of strategies available for p, while S_{qp} denotes a set of strategies that the player p assumes that q can prepare. That is, p believes that q's strategy set is S_{qp}. $\succeq_p$ denotes p's preference ordering on $S_p \times S_{qp}$, while $\succeq_{qp}$ is a preference ordering on $Sp \times S_{qp}$, which p assumes q to hold. That is, p supposes that q's preference ordering is $\succeq_{qp}$. We similarly define G_q. We naturally assume that $S_{ii} = S_i$ and $\succeq_{ii} = \succeq_i$ for $i = p$ and q. We also assume all the preference orderings are linear orderings and can be represented by some ordinal utility functions.

In this case G_p and G_q represent internal models of p and q, respectively, since they describe p's and q's understanding of the decision situation.

One of the most natural ways to describe the rational behavior of p and q in a simple hypergame may be obtained by modifying the Nash criterion for G_p and G_q in the following manner.

Let (G_p, G_q) be a simple hypergame with players p and q. Let

$$\mathcal{G}_p = \{G_p = (S_p, S_{qp}, \succeq_p, \succeq_{qp}) | S_p \subset \mathcal{S}_p, S_{qp} \subset \mathcal{S}_{qp},$$

$$\succeq_p \subset (S_p \times S_{qp}) \times (S_p \times S_{qp}), \succeq_{qp} \subset (S_p \times S_{qp}) \times (S_p \times S_{qp})\},$$

where $\mathcal{S}_p$ and $\mathcal{S}_{qp}$ are given supersets of S_p and S_{qp}, respectively.

We can define $\mathcal{G}_q$ in a symmetric way.

Definition 3.4 For a given simple hypergame (G_p, G_q), the Nash criterion for p is a function N_p from $\mathcal{G}_p$ into $\mathcal{P}(S_p \times S_{qp})$ such that for $G_p = (S_p, S_{qp}, \succeq_p, \succeq_{qp})$ in $\mathcal{G}_p$, $s^* = (s_p^*, s_{qp}^*)$ is in $N_p(G_p)$ if and only if we have

1. $s^* = (s_p^*, s_{qp}^*)$ is in $S_p \times S_{qp}$.
2. $(\forall s_p \in S_p)((s_p^*, s_{qp}^*) \succeq_p (s_p, s_{qp}^*))$ and $(\forall s_{qp} \in S_{qp})((s_p^*, s_{qp}^*) \succeq_q (s_p^*, s_{qp}))$ hold.

An element of $N_p(G_p)$ is called a Nash equilibrium of G_p. The definition claims: If $s^* = (s_p^*, s_{qp}^*)$ is a Nash equilibrium of G_p, then p believes that there is no incentive for either of the players to change their strategy as long as the other does not change its strategy. By applying symmetric arguments to $\mathcal{G}_q$, we can define $N_q : \mathcal{G}_q \to \mathcal{P}(S_{pq} \times S_q)$.

We now introduce overall rationality to deal with a simple hypergame, i.e., hyper-Nash equilibrium.

Definition 3.5 Hyper-Nash criterion is a function $K : \mathcal{G}_p \times \mathcal{G}_q \to \mathcal{P}(S_p \times S_q)$ such that for a simple hypergame (G_p, G_q) in $\mathcal{G}_p \times \mathcal{G}_q$, $s^* = (s_p^*, s_q^*)$ is in $K(G_p, G_q)$ if and only if

1. $s^* = (s_p^*, s_q^*)$ is in $S_p \times S_q$.
2. There are $s_{qp} \in S_{qp}$ and $s_{pq} \in S_{pq}$ such that (s_p^*, s_{qp}) is in $N_p(G_p)$ and (s_{pq}, s_q^*) is in $N_q(G_q)$.

$K(G_p, G_q)$ is a set of all pairs of strategies that are perceived as Nash equilibrium by at least one player. It is true that $K(G_p, G_q)$ may be empty or contain more than one element. An element of $K(G_p, G_q)$ is called a hyper-Nash equilibrium.

We may describe the decision situation above by $((G_p), N_p), (G_q, N_q); K)$ and call it a simple hypergame decision situation. It should be noticed that no explicit interaction is assumed between G_p and G_q, i.e., each player manages its perceived problem independently by using the Nash equilibrium criterion.

7.3.3 Symbiotic Hypergame

MLPM assumes that as time passes the players may begin to understand the other's interpretation of the situation. One of the ways to describe this is a symbiotic hypergame that is defined by simple hypergames and functions that represent how each player interprets the other's game. In the symbiotic hypergame, the players understand that they are concerned with a common situation, but they allow for different ways of identifying the situation.

Definition 3.6 A symbiotic hypergame with players p and q is a pair $((G_p, f), (G_q, g))$, where we have $G_p = (S_p, S_{qp}, \succeq_p, \succeq_{qp})$ and $f : S_q \rightarrow S_{qp}$, while $G_q = (S_{pq}, S_q, \succeq_{pq}, \succeq_q)$ and $g : S_p \rightarrow S_{pq}$ hold.

In $((G_p, f), (G_q, g))$ (G_p, G_q) is a simple hypergame defined by Definition 3.3 while the function $f : S_q \rightarrow S_{qp}$ represents how p interprets the set S_q of strategies of q. g has a symmetric interpretation. In this case (G_p, f) and (G_q, g) are internal models of p and q, respectively.

The following is a natural and straightforward way to define overall rationality for dealing with $((G_p, f), (G_q, g))$.

Definition 3.7 Let $((G_p, f), (G_q, g))$ be a symbiotic hypergame, where $G_p = (S_p, S_{qp}, \succeq_p, \succeq_{qp})$ and $G_q = (S_{pq}, S_q, \succeq_{pq}, \succeq_q)$ while $f : S_q \rightarrow S_{qp}$ and $g : S_p \rightarrow S_{pq}$. The symbiotic Nash criterion is a function L, which takes $((G_p, f), (G_q, g))$ to an element of $\mathcal{P}(S_p \times S_q)$ in such a way that (s_p^*, s_q^*) is in $L((G_p, f), (G_q, g))$ if and only if

1. $(s_p^*, s_q^*) \in S_p \times S_q$
2. $(s_p^*, f(s_q^*)) \in N_p(G_p)$ and $(g(s_p^*), s_q^*) \in N_q(G_q)$.

We call an element of $L((G_p, f), (G_q, g))$ a symbiotic Nash equilibrium of $((G_p, f), (G_q, g))$.

Let $(s_p^*, s_q^*) \in S_p \times S_q$ be a symbiotic Nash equilibrium of $((G_p, f), (G_q, g))$. Then, player p, who perceives it as $(s_p^*, f(s_q^*)) \in N_p(G_p)$ has no incentive to change its strategy from s_p^* as long as p believes that q will not change its strategy from $f(s_q^*)$. Since a similar argument holds for q, a symbiotic Nash equilibrium can be seen as a natural extension of the Nash equilibrium by taking into account interpretation by the players. The decision situation above could be characterized by $((G_p, f), N_p), ((G_q, g), N_q); L)$.

7.3.4 Hypergame Sharing the Same Value System

As time goes on, understanding of the situation by the players may become shared by both players and the symbiotic hypergame may produce a sort of consistency between the interpretations. We will define consistency as follows:

Definition 3.8 Let $((G_p, f), (G_q, g))$ be a symbiotic hypergame. We say that p perceives q's preference with global consistency (with respect to (f, g)) if for any $s_p \in S_p$ and any s_q and $s'_q \in S_q$ we have

$$(g(s_p), s_q) \succeq_q (g(s_p), s'_q) \Leftrightarrow (s_p, f(s_q)) \succeq_{qp} (s_p, f(s'_q)).$$

The definition claims that p's perception $\succeq_{qp}$ of q's preference is consistent with respect to f and g. Similarly, we have a symmetric definition for q.

Definition 3.9 Symbiotic hypergame $((G_p, f), (G_q, g))$ is said to share the same value system if p perceives q's preference with global consistency and q perceives p's preference with global consistency, i.e., if we have

$$(\forall s_p \in S_p)(\forall s_q, s'_q \in S_q)(g(s_p), s_q) \succeq_q (g(s_p), s'_q) \Leftrightarrow (s_p, f(s_q)) \succeq_{qp} (s_p, f(s'_q))$$

and

$$(\forall s_p, s'_p \in S_p)(\forall s_q \in S_q)((s_p, f(s_q)) \succeq_p (s'_p, f(s_q)) \Leftrightarrow (g(s_p), s_q)) \succeq_{pq} (g(s'_p), s_q)).$$

As more time passes, the hypergame sharing the same value system may change to a usual noncooperative game where both the players face the same game $G = (S_p, S_q, \succeq_p, \succeq_q)$. This common perception may be broken up intentionally or unintentionally and a new problematic situation may arise.

The four stages mentioned above can be illustrated by Fig. 7.5. It also illustrates that the game situations do not necessarily follow the outer cycle but may jump from one stage to another taking the straight lines inside.

Since MLPM assumes a problematic situation to change structurally and spontaneously along the learning process, it is, in principle, impossible to explain how each player follows the process a priori: The only way to do it may be to introduce a higher mechanism or model a priori that controls the process of MLPM in such a way that parameters of the higher mechanism characterize each stage of MLPM. However, here we face a serious problem: How to explain the control of the higher mechanism itself. To control it we need a second higher mechanism, which controls the controller of the process of MLPM. In this way, we need infinitely many hierarchies, where each level controls the lower level (van Gigch, 1991).

In this article, however, we do not introduce such a mechanism but will only focus on investigating how equilibria disappear or emerge along the process: We will try to find, for example, under what conditions they disappear with $((G_p, f), (G_q, g))$ even though there are some equilibria of G_p or G_q, or conversely, under what conditions the learning process generates new equilibria of $((G_p, f), (G_q, g))$ though there is no equilibrium of G_p or G_q. Since these kinds of generation and disappearance problems of equilibria are essentially concerned with "emergent properties" of the process, to find conditions under which the phenomena happen is very meaningful from a systems theoretic point of view (Checkland, 1981; Mesarovic and Takahara, 1989). Furthermore, if we can obtain such conditions, we will be able to reveal under what conditions what phenomena will happen (what-if analysis).

Proposition 3.1 (Kijima, 1996) *Let* $(G_p, G_q))$ *and* $((G_p, f), (G_q, g))$ *be a simple hypergame and hypergame sharing the same value system, respectively. If* f *and* g *are surjective, we have*

$$N_p(G_p) \neq \phi \Leftrightarrow N_q(G_q) \neq \phi \Leftrightarrow L((G_p, f), (G_q, g)) \neq \phi.$$

$f : S_q \to S_{qp}$ is called surjective if we have $(\forall y \in S_{qp})(\exists x \in S_q)(f(x) = y)$.

Proof

(1) $N_p(G_p) \neq \phi \Leftrightarrow N_q(G_q) \neq \phi$:

Suppose $N_p(G_p) \neq \phi$, then there is $(s_p^*, s_{qp}^*) \in N_p(G_p)$. Since f is surjective, for s_{qp}^* there is $u^* \in S_q$ such that $s_{qp}^* = f(u^*)$. From the definition of hyper-Nash equilibrium, we have $(\forall s_p \in S_p)((s_p^*, f(u^*)) \succeq_p (s_p, f(u^*))$ and $(\forall s_{qp} \in S_{qp})(s_p^*, f(u^*)) \succeq_{qp} (s_p^*, s_{qp}))$.

The latter immediately implies $(\forall s_q \in S_q)(s_p^*, f(u^*)) \succeq_{qp} (s_p^*, f(s_q)))$ since for each $s_q \in S_q$ we have $f(s_q) \in S_{qp}$. It follows from the global consistency of p that $(\forall s_q \in S_q)((g(s_p^*), u^*) \succeq_q (g(s_p^*), s_q))$.

Similarly, we have $(\forall s_{pq} \in S_{pq})((g(s_p^*), u^*) \succeq_{pq} (s_{pq}, u^*))$. Consequently, we have $(g(s_p^*), u^*) \in S_{pq} \times S_q$ is a hyper-Nash equilibrium of G_q, i.e., $N_q(G_q) \neq \phi$.

To show the opposite implication we can apply a similar argument.

(2) $N_p(G_p) \neq \phi \Leftrightarrow L((G_p, f), (G_q, g) \neq \phi$:

$(\Rightarrow)$: Assume that $N_p(G_p) \neq \phi$ and $L((G_p, f), (G_q, g) = \phi$. Let (x^*, y^*) be in $N_p(G_p)$. Then, since f is surjective, there is $y^0 \in S_q$ such that $y^* = f(y^0)$.

It follows from $(x^*, y^*) \in N_p(G_p)$ that

$$(\forall x \in S_p)((x^*, f(y^0)) \succeq_p (x, f(y^0)) \tag{7.1}$$

and

$$(\forall y' \in S_{qp})((x^*, f(y^0)) \succeq_{qp} (x^*, y')). \tag{7.2}$$

Equation (7.2) implies

$$(\forall y \in S_q)((x^*, f(y^0)) \succeq_{qp} (x^*, f(y))) \tag{7.3}$$

since $(\forall y \in S_{qp})(f(y) \in S_{qp})$ holds.

Because of the global consistency, (7.1) and (7.3) imply

$$(\forall x \in S_p)((g(x^*), y^0) \succeq_{pq} (g(x), y^0)) \tag{7.4}$$

and

$$(\forall y \in S_q)((g(x^*), y^0) \succeq_q (g(x^*), y)). \tag{7.5}$$

Since g is surjective, Eq. (7.4) implies

$$(\forall x' \in S_{pq})((g(x^*), y^0) \succeq_{pq} (x', y^0)). \tag{7.6}$$

On the other hand, it follows from $L((G_p, f), (G_q, g) = \phi$ that (x^*, y^0) cannot be a symbiotic Nash equilibrium, i.e., we should have $(x^*, f(y^0)) \notin N_p(G_p))$ or $(g(x^*), y^0) \notin N_q(G_q)$. However, since $(x*, f(y^0)) \in N_p(G_p)$ holds, we must have $(g(x^*), y^0) \notin N_q(G_q)$.

It follows from the definition of $N_q(G_q)$ that

$$(\exists x' \in S_{pq})((g(x^*), y^0) \prec_{pq} (x', y^0)) \tag{7.7}$$

or

$$(\exists y \in S_q)((g(x^*), y^0) \prec_q (g(x^*), y)). \tag{7.8}$$

However, Eq. (7.7) contradicts Eq. (7.6) while Eq. (7.5) contradicts Eq. (7.8).

($\Leftarrow$): Assume that $L((G_p, f), (G_q, g) \neq \phi$ and$N_p(G_p) = \phi$. Let (x^*, y^*) be in $L((G_p, f), (G_q, g))$. By definition we have $(x^*, f(y^*)) \in N_p(G_p)$ and $(g(x^*), y^*) \in N_q(G_q)$. However, the former contradicts the assumption $N_p(G_p) = \phi$. Q.E.D.

A similar argument is applicable to g. The condition of the proposition that f is surjective intuitively means that p estimates q's strategy set S_q without any redundant strategies.

Hence, we can interpret the preposition in the following way: Suppose at first there are two independent G_p and G_q, and now there emerge interpretation functions f and g between them. Assume the players with f and g are such that they share the same value system and each estimates the other's strategy set without redundancy. Then, there is a Nash equilibrium for p if and only if there is a Nash equilibrium for q if and only if there is a symbiotic Nash equilibrium of $((G_p, f), (G_q, g))$. That is, solutions of p and q do not disappear by generation of the interpretation functions f and g.

7.4 Application of MLPM to the Gulf War

In this section we will apply MLPM developed theoretically so far to an actual and dynamic problematic situation, namely, the Gulf War (Kijima, 1996). In this application we try to demonstrate how the model works and what insights into the situation we could obtain. We construct the players' internal model based on published data as realistic as possible, although our emphasis is more on explaining how MLPM dynamically analyzes a soft problematic situation: We will show that when MLPM is adopted, it allows us to look at the problematic situation involving the players from high up so that we can obtain somewhat deeper insights into the situation.

The Gulf War, which began on January 17, 1991, was a large-scale war between multinational military forces (of which the main part was provided by the USA) and Iraq (Okakura, 1991). The war is usually divided into three periods, but in this application we will focus mainly on the second period. It is because during this period the evaluation and perception of the strategies by both parties changed so drastically that we believe analysis of this period by MLPM should illustrate its essence.

The First Period (from August 2, 1990 to August 8, 1990) The invasion of Kuwait by the Iraqi army on August 2, 1990 triggered the following large-scale conflict. On August 8 Iraq had completed the invasion and declared that it had annexed Kuwait. Responding to this declaration, the President of the USA. officially announced the decision to deploy the US military force to Saudi Arabia. Although until then Iraq had not expected a military fight against the USA, the announcement changed their identification of the situation. At the same time, the announcement opened the door for the USA to take strong military measures and led to high tension between the two countries.

The Second Period (from August 9, 1990 to January 9, 1991) Backed by a huge military force, the USA demanded Iraq to withdraw from Kuwait unconditionally, though Iraq struggled to seek a solution by linking the issue to other international problems such as the Palestinian–Israel conflict. To resolve the confrontation the summit talks between the two countries took place on January 9, 1991, which was just a few days before the deadline for withdrawal, January 15, 1991, set by the United Nations. However, the talks failed to reach an agreement and there seemed to be no room for any outcome except war.

The Third Period (from January 10, 1991 to January 17, 1991) Finally, the parties went to war on January 17, 1991.

In this analysis the USA (A) and Iraq (I) are clearly the players of the game.

7.4.1 Simple Hypergame Analysis

We represent an earlier stage of the period by a simple hypergame (G_A, G_I), where $G_A = (S_A, S_{IA}, \succeq_A, \succeq_{IA})$ is the USA's subjective game (Table 7.1), while $G_I = (S_{AI}, S_I, \succeq_{AI}, \succeq_I)$ is Iraq's subjective game (Table 7.2).

In Table 7.1 G_A shows the set of strategies available for the USA is $S_A = \{F, MP, E\}$. F denotes more *Flexible* treatment of the situation than before such as reduction of military pressure and reliance mainly on economic sanctions to Iraq. MP means continuation of *Military Pressure* over Iraq in a similar manner as in the first period, while E stands for *Escalation* of military action.

G_A also represents that the USA believed the set of strategies available for Iraq as $S_{IA} = \{UW, CW, H\}$. UW means *Unconditional Withdrawal*, while CW denotes *Conditional Withdrawal*. H stands for *Hostile* action against the USA.

The left side of the payoff of G_A indicates that the USA was willing to demand Iraq to withdraw unconditionally. If Iraq conceded, the USA was most willing to handle the situation as moderately as it could. However, if Iraq took other actions, the USA got prepared to deal with Iraq by escalating the military force, taking strict measures and refusing any demands by Iraq. The right side of the payoff of G_A represents the USA believed that although Iraq was basically continuing its hostile attitude toward the USA, it, at the same time, was also ready for conditional withdrawal if the USA took a flexible option. Furthermore, the USA believed that Iraq would withdraw if the USA took the military option.

In Table 7.2 G_I shows the set of Iraq's strategies was $S_I = \{SW, N, A\}$. SW stands for *Strategic Withdrawal*, which means withdrawal with some gain like the occupation of some portion of the petroleum wells in Kuwait. N means *Negotiation* for resolution while A stands for *Aggression* of continuation of the invasion.

G_I indicates Iraq believed the set of strategies of the USA as $\{P, NP\}$, i.e., $S_{AI} = \{P, NP\}$. P is *Provocative* actions against Iraq, while NP represents *Non-Provocative* options.

The right side of the payoff of G_I reflects Iraq's basic attitude of legitimacy of the invasion against any options the USA intended to resolve the conflict. Iraq tried to avoid unconditional withdrawal at any cost. The left side of the payoff of G_I implies

Table 7.1 G_A: Subjective game of USA

		S_{IA}		
		UW	CW	H
	F	9 5	2 9	1 6
S_A	MP	8 2	4 4	3 8
	E	7 1	6 3	5 7

Table 7.2 G_I: Subjective game of Iraq

		S_I		
		SW	N	A
S_{AI}	NP	6 2	5 6	2 4
	P	4 1	3 5	1 3

Iraq believed that though the USA felt strong frustration against the continuation of the invasion by Iraq, it would take non-provocative actions in response to other strategies of Iraq.

By conducting simple hypergame analysis to (G_A, G_I), we find $N_A(G_A) = \{(E, H)\}$ and $N_I(G_I) = \{(NP, N)\}$, respectively. Hence, we have $K(G_A, G_I) = \{(E, N)\}$, as the overall rational outcome. That is, the USA escalates military action and Iraq seeks for negotiation for resolution.

7.4.2 Symbiotic Hypergame Analysis

Now we will discuss how the equilibrium of (G_A, G_I) would change as mutual learning between the players unfolded. Suppose the mutual learning is represented by interpretation functions defined by

$$
\begin{aligned}
f : S_I &\rightarrow S_{IA} \\
SW &\mapsto CW \\
N &\mapsto H \\
A &\mapsto H
\end{aligned}
$$

$$
\begin{aligned}
g : S_A &\rightarrow S_{AI} \\
F &\mapsto NP \\
MP &\mapsto P \\
E &\mapsto P
\end{aligned}
$$

That is, f shows the USA identified strategic withdrawal (SW) with conditional withdrawal, and regarded both the request for negotiation (N) and continuation of the invasion (A) as hostile options. Iraq has no strategy to be interpreted as unconditional withdrawal (UW) by the USA; f is not surjective. g represents that Iraq considered only the flexible option (F) of the USA as non-provocative (NP), while military pressure (MP) and its escalation (E) were seen as provocative (P).

The USA interpreted $(E, N) \in K(G_A, G_I)$ as $(E, f(N)) = (E, H)$, which is in $N_A(G_A)$. It implies that the USA understood (E, N) as its Nash equilibrium. This interpretation of (E, N) by the USA led to military actions by the USA: On November 8, 1990, to demonstrate military presence in the area, the USA deployed around two hundred thousand additional soldiers to Saudi Arabia. However, (E, N) is interpreted as $(g(E), N) = (P, N)$ by Iraq and it is not in $N_I(G_I)$) so that Iraq recognized its subjective game might be incorrect and revised its preference $\succeq_{AI}$ to $\succeq'_{AI}$ and redefined its game as G'_I given by Table 7.3.

Table 7.3 G'_I: Revised subjective game of Iraq

		S_I		
		SW	N	A
S_{AI}	NP	3 2	2 6	1 4
	P	6 1	5 5	4 3

Table 7.4 $\hat{G}_A$: What-if analysis

		S_{IA}		
		UW	CW	H
S_A	F	9 5	7 9	1 6
	MP	8 2	6 4	2 8
	E	5 1	4 3	3 7

By applying MLPM to G'_I, we have $N_I(G'_I) = \{(P, N)\}$. This means that Iraq supposed that the USA would take P while Iraq kept N unchanged as its own strategy. At this moment it follows from $N_A(G_A) = \{(E, H)\}$ and $N_I(G'_I) = \{(P, N)\}$ that $K(G_A, G'_I) = \{(E, N)\}$. This is interpreted as $(E, f(N)) = (E, H) \in N_A(G_A)$ by the USA and $(g(E), N) = (P, N) \in N_I(G'_I)$ by Iraq. Hence both the USA and Iraq recognized it as their own equilibrium, i.e., $(E, N) \in L^*((G_A, f), (G'_I, g))$.

Consequently the outcome was settled at (E, N) and the summit talks between the USA and Iraq took place. However, the interpretation of (E, N) as (E, H) by the USA implies that it understood the talks as opportunities to explain why the military escalation was legitimate as well as to demand Iraq to withdraw unconditionally. On the other hand, Iraq's interpretation of (E, N) as (P, N) shows that Iraq understood the talks as occasions to avoid military fights in the most preferable way, since Iraq had sought negotiation opportunities by approaching the U.S.S.R. and France. Actually the summit talks, due to the gap between the parties' basic attitudes, could not resolve the conflict in a peaceful way.

7.4.3 What-if Analysis

So far, by using MLPM we have analyzed the changing process of identification of the war from the viewpoints of both the players to provide rational logic behind their behavior. We could find two key factors in the USA's game, which led the USA to the war, i.e., the payoff shown by G_A and the shape of the interpretation function f.

To conduct what-if analysis, now let us suppose that the USA seriously considered Iraq's request for negotiation by introducing a slightly different game $\hat{G}_A$ and $\hat{f} : S_I \to S_{IA}$.

$\hat{G}_A$ is given by Table 7.4 and $\hat{f}$ is defined by

$$\hat{f} : S_I \to S_{IA}$$
$$SW \mapsto CW$$

$$N \mapsto CW$$

$$A \mapsto H$$

It means that the USA is assumed to consider that even if Iraq withdraws *conditionally* it would take the flexible attitude toward Iraq.

Then, $N_A(\hat{G}_A) = \{(F, CW), (E, H)\}$. If the game seen by Iraq is assumed to remain the same as G_I, we still have $N_I(G_I) = \{(NP, N)\}$ and consequently $K(\hat{G}_A, G_I)$ should be $\{(F, N), (E, N)\}$. (F, N) is interpreted by the USA as $(F, \hat{f}(N)) = (F, CW)$, which is in $N_A(\hat{G}_A)$, while it is by Iraq as $(g(F), N) = (NP, N)$, which is also in $N_I(G_I)$. Hence, both the players understand they have realized (F, N) in $L^*((\hat{G}_A, \hat{f}), (G_I, g))$.

The what-if analysis implies that if they have symbiotic hypergame $((\hat{G}_A, \hat{f}), (G_I, g))$, not only (E, N) but also (F, N) have possibility to occur. Since (F, N) implies the USA takes the flexible attitude while Iraq intends to negotiate, they could avoid the military confrontation.

7.5 Conclusion

The present article investigates mutual learning process of two players in a soft problematic situation to propose MLPM characterized by four stages, i.e., isolated, mutual interpretation, value sharing and complete mutual understanding stages, and transitions among them. We distinguish among the stages in terms of internal models and rationality of the players. We also examine how emergent properties arise along the learning process. Furthermore, we applied it to the Gulf War, to gain unique insights into the dynamic change of the players' attitude toward the war. We claim that the model could provide a new way for investigating soft problematic situations theoretically and practically. For example, when we formulate several wars in the same framework of MLPM, some commonalities in the metrics or interpretation functions among them may allow us to transfer some findings about a war to other wars.

References

Bennett, P. G. (1980). Hypergames: Developing a model of conflict. *Futures*, *12*(6), 489–507.

Bennett, P. G., & Dando, M. R. (1979), Complex strategic analysis: A hypergame study of the fall of France. *Journal of Operational Research Society*, *30*(1), 23–32.

Bennett, P. G., Dando, M. R., & Sharp, R. G. (1980). Using hypergames to model difficult social issues: An approach to the case of soccer hooliganism. *Journal of Operational Research Society*, *31*(7), 621–635.

Bennett, P. G., Cropper, S., & Huxham, C. (1989). Modelling interactive decisions: The hypergame focus. In Rosenhead, J. (Ed.), *Rational analysis for a problematic world*. John Wiley and Sons.

Bryant, J. W. (2015). *Acting strategically using drama theory*. CRC Press.

Checkland, P. B. (1981). *Systems thinking, systems practice*. John Wiley.
Gibbons, R. (1992). *A primer in game theory*. Harvester Wheatsheaf.
Kijima, K. (1996). An intelligent poly-agent learning model and its application. *Information and Systems Engineering*, *2*, 47–61.
Kovach, N. S., Gibson, A. S., & Lamont Kovach, G. B. (2015). Hypergame theory: a model for conflict, misperception, and deception. *Game Theory*, *2015*, Article ID 570639. https://doi.org/10.1155/2015/570639
Mesarovic, M. D., & Takahara, Y. (1989). *Abstract systems theory*. Springer.
Neches, R., et al. (1991). Enabling technology for knowledge sharing. *AI Magazine*, *12*(3), 36–56.
Okakura, T. (1991). The Gulf crisis. In *Sekai*, Iwanami, Tokyo (in Japanese).
Rosenhead, J. (Ed.). (1989). *Rational analysis for a problematic world*. John Wiley and Sons.
van Gigch, J. P. (1991). *System design modeling and metamodeling*. Plenum Press.
Wonham, W. N. (1976). Towards an abstract internal model principle. *IEEE Transactions on Systems, Man, and Cybernetics*, *6*(11), 735–740.

Chapter 8
Model Theory Approach for Simulation: Improvements of Model Description Language and Integration of Development Environments

Takao Asahi

Abstract Model Theory Approach for Simulation is an application of mathematical systems theory to the development of simulations. This approach has the following characteristic. (1) System models are defined by the well-formed formula in first-order logic. (2) The mathematical models are coded in computer language CAST (Computer Acceptable Set Theory). (3) The coded models are compiled by MTA (System Development Environment). (4) Finally, the compiled models are attached with a standard execution engine. Then you can get the working systems. This chapter shows the overview of the approach, some examples of simulation, and recent technical advancement.

Keywords Predicate logic · Automaton · Simulation · Model theory · Model description language CAST

8.1 Introduction

While constructing a mathematical general systems theory, we have been applied the results to various fields of Management Information Systems (MIS). It is called "Model Theory Approach" for MIS developments. We have already created a model description language CAST (Asahi et al., 2008) and development-execution environment (software) for information system development, and applied it to web-based transaction processing systems such as hotel room management systems including problem-solving systems or goal seeking systems (Takahara et al., 2004, 2007, 2008).

Since 2011, we have set a short-term research theme "construction of general models and simulation development methodology for autonomous distributed

T. Asahi (✉)
Toyo University, Tokyo, Japan
e-mail: asahi@toyo.jp

K. Kijima et al. (eds.), *Systems Research I*, Translational Systems Sciences 26,
https://doi.org/10.1007/978-981-16-9937-5_8

systems." And have improved development-execution environment and model description language for application in the simulation field. As a result, the structure of the mathematical model and the model by the language CAST became almost the same, so full-fledged simulation system development became possible. The purpose of this chapter is to outline the research results (theory and technology) so far; to report on the improvement of the model description language and the development-execution environment, and to clarify the significance from the viewpoint of the model theory approach.

Section 8.2 describes the features of information system development by model theory approach, especially from the viewpoint of the development procedure. On the other hand, another feature of the model theory approach is to have a background theory. In Sect. 8.3, we describe a method of simulation development based on theory. Section 8.4 describes the role of the newly added predicate (equivalent to commands) necessary for the consistency of theory and practice in CAST (Computer Acceptable Set Theory), a computer language for model description. Section 8.5 and 8.6 discuss the improved situation of the development-execution environment, which is software for compiling and executing models. Section 8.7 is a summary.

8.2 Model Theory Approach

The feature of information system development by model theory approach is to define a model of the target system by logical expression from the viewpoint of systems theory and to create a computational model by coding the mathematical model using the computer language CAST. It can be executed directly by compiling it in a system development-execution environment (Takahara et al., 2007, p. 5).

As for the "image of target system" in this development procedure (Fig. 8.1), we have been able to develop problem-solving systems (Solver), transaction processing

image of target system
↓ (systems theory)
Modeling
↓ (mathematical model)
Coding
↓ (language CAST)
user model (computational model)
↓ (development-execution environment)
Compiling
↓ (standard execution engine)
target system (working system)

Fig. 8.1 Information system development procedure in Model Theory Approach. (Revised from Asahi, 2011, p. 186)

systems (TPS), and their composite systems. On the development of simulation systems (Simulation), automaton systems and composite systems with problem-solving systems was added recently. The implementation structure of the model varies depending on the type of these target systems.

In addition, there are two types of "system models": mathematical models based on system theory and computational models based on language CAST (called "user model" in model theory approach). The "similarity" (or isomorphic structure) of these two kinds of models is critical in actual system development. It is necessary to improve the thought efficiency and development efficiency that both models are almost similar or the same type in the look. Because of its similarity, mathematical models will be no longer needed if we get used to it, and we can directly create computational models using the language CAST. The computational model (created as a text file) is called the "user model" in the model theory approach.

On importing the user model into the development-execution environment, it is compiled into a system that runs. But a program (called an execution engine) to execute the model is required separately at that time. However, in the model theory approach, since various general standard execution engines are prepared according to the type of target system described above, it is only necessary for the developer to specify which execution engine to add. In other words, developers do not need to create engines to run models and can focus only on modeling. This is also a powerful advantage of the model theory approach.

8.3 Background Theory and Simulation Development

Among the various field of MIS development, this chapter focuses on simulation development which is the recent research theme. This section explains the relationship between the background theory and the development procedure.

In simulation development using the model theory approach, we can develop a composite system of automata and problem-solving systems. Therefore, the background theory is a well-known standard theory of automaton and a theory of problem-solving. Note that we use only the Mealy-type model of automaton in the case of simulation development. In addition, since problem-solving is not dealt with in this chapter, please refer to Takahara et al. (2008).

First of all, the basic principle is that "a system in which multiple automata are connected is one automaton as a whole." In this chapter, we call it the principle of consistency. In addition, a system in which multiple automata are connected is called a composite system.

According to the principle of consistency, the development policy of the execution engine (software) for the user model is established as follows. That is, if there is an engine that can run one automaton, it should be able to run any composite system. This is the development policy for creating an execution engine. In fact, a standard execution engine called stdAutomatonEngine.p is added to the user model in compiling phase (Fig. 8.1).

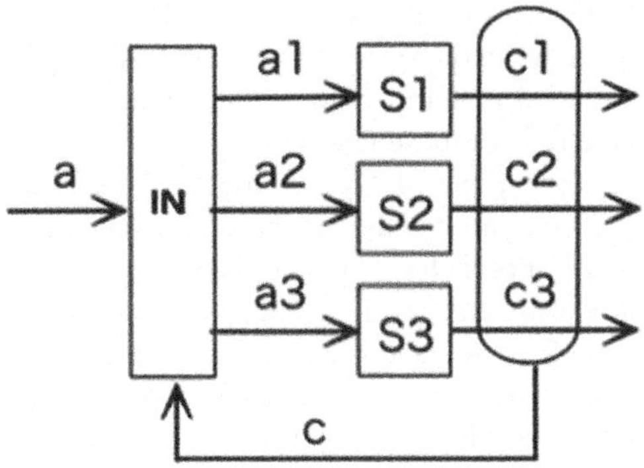

Fig. 8.2 General form of a concurrent composite system S of multiple state systems (Asahi, 2012, 2013)

The operation algorithm of the standard execution engine is very simple. Each automaton can be modeled by describing two functions (state transition function δ and output function λ). In the model theory approach, each function is described as a well-formed formula using logical expressions. Therefore, the standard execution engine should evaluate the truth value of these functions repeatedly (Asahi et al., 2008).

For example, Fig. 8.2 is a block diagram that shows three state systems S_1, S_2, and S_3 composed together. You can think of the feedback (input connection function IN: $A \times C_1 \times C_2 \times C_3 \rightarrow A_1 \times A_2 \times A_3$) as the interaction between the subsystems with each other. As for the model description, if you specify the input connection function IN appropriately, then three subsystems can be combined in series, parallel, and feedback connection, respectively.

When the state transition functions of each state system S_1, S_2, and S_3 are defined as δ_1, δ_2, and δ_3, respectively, the state transition function δ of the entire system in Fig. 8.2 can be defined as Eq. (8.1), where the symbol $\Longleftrightarrow$ in the equation is a necessary and sufficient condition (the left side is defined by the right side), and the comma at the right end of each line represents the logical connection "and." Note that the symbol "cc" is not multiplication but a single variable. Equation (8.1) is called a well-formed formula in the field of logic. We call it as "mathematical model of the system S."

$$\begin{aligned} \delta\,(c, a) = \mathrm{cc} \Longleftrightarrow & \\ & (c_1, c_2, c_3) = c, \\ & (a_1, a_2, a_3) = \mathrm{IN}\,(a, c)\,, \\ & \mathrm{cc} = (\delta_1\,(c_1, a_1)\,, \delta_2\,(c_2, a_2)\,, \delta_3\,(c_3, a_3))\,. \end{aligned} \tag{8.1}$$

Next, Eq. (8.2) is a user model ("computational model") of the entire system S of Fig. 8.2, where the symbol ":=" means that the result of the operation on the right side should be substituted into the variable on the left side.

$$\begin{aligned} \text{delta}\,(c, a) = \text{cc} <-> \\ &(c1, c2, c3) := c, \\ &(a1, a2, a3) := \text{IN}\,(a, c)\,, \\ &\text{cc} := (\text{delta1}\,(c1, a1)\,, \text{delta2}\,(c2, a2)\,, \text{delta3}\,(c3, a3))\,. \end{aligned} \tag{8.2}$$

In practice, it is important that the structure and items of mathematical models such as Eq. (8.1) be coded in the language CAST as directly as possible, which increases the efficiency of development. Even if the number of subsystems increases and the coupling becomes complicated in a system, a model description can be performed with a consistent structure (parallel arrangement + feedback coupling). It can be said that is a principle of model description.

Note that a composite system that performs sequential processing and concurrent processing has a different model as a whole. For sequential processing, as usual, the next process is coded on the next line. On the other hand, with regard to concurrent processing, the principle that "a composite system that performs concurrent processing can be modeled in a simple form (parallel placement + feedback coupling in Fig. 8.2)" is important. This is the standard form of a concurrent processing system whose subsystems are state systems (Asahi, 2012, 2013).

On the other hand, the model development method will follow the principle of consistency. A medium-sized composite model is composed by combining multiple elemental automata, and a multilayer system is assembled by combining multiple medium-sized models again. Such step-by-step modeling can be consistently supported by the same principle. As a result, it is possible to simulate complexly coupled systems.

The above principles can be proven mathematically too. In addition to applying the results of system theories to practice, we will assemble background theories necessary for simulation development. In other words, it can be said that practice enriches the theory. That is, it is an interaction between theory and practice.

8.4 Model Description Language CAST

Until 2010, we have developed basic technologies with an emphasis on running anyway, being able to draw, and being able to develop many types of simulation systems (or showing possibilities). However, in the model description language CAST, there were still problems in the consistency of theory and practice and in the speedup of execution.

On the consistency between theory and practice, there was a problem of isomorphism (structural similarity) between mathematical models and user models by language CAST. In order for both models to be isomorphic, it is necessary to be flexible enough that the language CAST can describe mathematical models as mentioned in Sect. 8.3.

Already, problem-solving systems and transaction processing systems could be developed by the model theory approach (Takahara et al., 2007). But in the application to simulation, it remained left how to describe many objects, and how to read and write efficiently the values of spreadsheets.

8.4.1 Description of Many Items

At the modeling stage, "subscript" is often required for describing many items (e.g., variables x_i, functions f_i, and sets S_i). In addition, if we have a simple format as much as possible for defining vectors or lists with many items, the visibility and listability will also increase, and the efficiency of development will follow.

For the former, variables and sets with subscripts are defined as functions in the language CAST. For example, variables x_i and sets S_i can be defined as functions $x(i)$ and $S(i)$, respectively. And at a glance, the two-variable functions $\delta_i(c_i, a_i)$ should be defined as the three variable function delta(i, c_i, a_i) in the language CAST, but it could not so far. The reason is that the names of the main functions were constrained as reserved words. However, from now on, it will be possible to define state transition functions and output functions with subscripts without the need for function-declaration "func([])". It is convenient (Asahi, 2015).

As for the latter, there has been a method of using the list definition function "defList" as a method of describing vectors or lists with many items in the language CAST. However, since its usage was not clear, an accurate mathematical definition of defList was given. And a reliable defList usage for many item descriptions was established (Asahi, 2014). Intuitively speaking, defList is a function that lists the first value of y such that, for element x of index list Xs, y is the first value to determine the list configuration-condition $p(y, x)$ to be true.

Thanks to the former subscripts and the latter function defList, it is allowed, for example, to describe the state transition function delta(c, a) of the entire system with only two logical expressions as next Eqs. (8.3) and (8.4). Even if the number "Ns.g" of subsystems in Fig. 8.2 increased, the form is the same (Asahi, 2015).

$$\text{delta}\,(c, a) = \text{cc} <-> \text{cc} := \text{defList}\,(p\,(\text{cci}, i, [a, c])\,, \text{member}\,(i, \text{Ns.g}))\,, \tag{8.3}$$

$$p\,(\text{cci}, i, [a, c]) <-> \text{cci} = \text{delta}\,(i, \text{project}\,(c, i)\,, \text{IN}\,(i, a, c))\,. \tag{8.4}$$

This is an expanded definition of Eq. (8.2). Indeed, if the index list is defined as Ns.g: = [1, 2, 3], then Eqs. (8.3) and (8.4) become the same as Eq. (8.2), because project(c, i) = c_i and IN(i, a, c) = a_i for $i = 1, 2, 3$.

In other words, even if the number of subsystems in a system increased and the coupling became complicated, the description of the many-component system became consistently the same format (two logical expressions).

8.4.2 "Read and Write" of Spreadsheet

There was already a predicate (equivalent to a command) for memorizing (and referring to) the state of each system into a spreadsheet, but there was no description method equivalent to the mathematical model. In order to maintain the isomorphic structure and to simplify as much as possible, we have defined and added a new predicate "cell(wp, x, y)" at this time. Where, "cell(wp, x, y)" means the value of a cell in the row x column y of the sheet number wp. As a result, for example, the following definition can be made in the language CAST (Asahi, 2015).

$$\text{delta}(a) <--> \text{cell}\,(\text{wp}, 1, 1) := \text{delta}\,(\text{cell}\,(\text{wp}, 1, 1)\,, a)\,. \tag{8.5}$$

This Eq. (8.5) defines the state transition of a system as (not a function) but a predicate delta(a), where the current value of state is assumed to be in the cell(wp, 1, 1). Using the current input "a" and the current state value obtained from cell(wp, 1, 1), the predicate delta(cell(wp, 1, 1), a) calculates the next state. The result is overwritten to cell(wp, 1, 1).

8.4.3 High-Speed Execution of Many-Element Systems

On the other hand, thanks to spreadsheet, the speedup of simulation execution have been made available. In other words, instead of storing the state of a system in a variable in the model, it is possible to speed it up by storing it outside the model.

The idea of physical realization is described on a method of realizing automaton as a "physical system." In mathematical models, a normal automaton is usually considered a dynamic system. But the state transition function is realized as a function (static system), and the storage is realized by using a storage device which is a dynamic system, and the expected system is realized by combining them. The physical realization system is also an automaton and is consistent with the behavior of the original system.

Asahi (2015) gave the physical realization of multiple-element systems (Fig. 8.3). This makes us construct mathematical models of the system that stores the value of states outside the model. Moreover, by using the above results (8.1) and (8.2), it is possible to describe a model in the same structure using the language CAST. For example, if you want a spreadsheet to remember the state of your system (i.e., the storage in Fig. 8.3 is a spreadsheet), write the model as follows:

$$\text{delta}\ \ P(a) <--> \text{Zs} := \text{defList}\,(p\,(z, i, [a])\,, \text{member}\,(i, \text{Ns.g}))\,, \tag{8.6}$$

$$p\,(z, i, [a]) <--> \text{cell}\,(\text{wp2.g}, 1, i) := \text{delta}\,(i, \text{cell}\,(\text{wp1.g}, 1, i)\,, \text{IN}\,(i, a))\,. \tag{8.7}$$

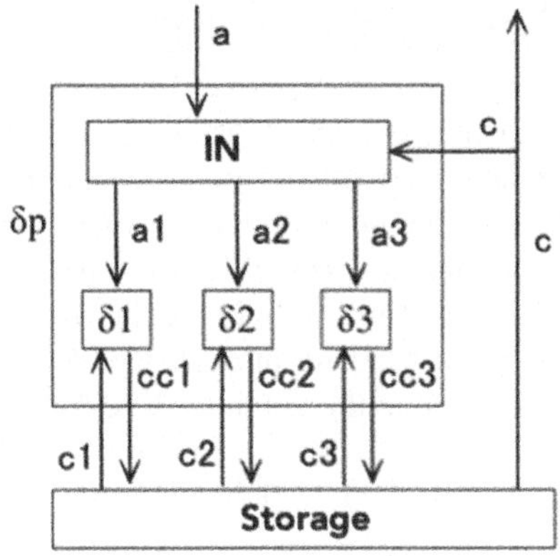

Fig. 8.3 Physical realization of Fig. 8.2 with three subsystems (Asahi, 2015)

That is, even if the target system is a complex system with many elements, its model in language CAST can be constructed in almost the same structure as Eq. (8.3) and (8.4). And as a by-product, it's now possible to run faster.

8.5 Development-Execution Environment (New MTA-SDK)

In this section, we describe the current status of the development-execution environment (software) for compiling the system model and creating it into a running system.

8.5.1 Integration of Development-Execution Environment

So far, the development of the development-execution environment for simulation has been carried out under the name Simcast separately from the development-execution environment for conventional information systems, MTA-SDK (System Development Kit). But both environments have been integrated (Takahara et al., 2016).

The development infrastructure of the problem-solving system which is the basis of artificial intelligence, and the web-based transaction processing system have already been incorporated into MTA-SDK. A database query language SQL and Simcast for simulation were newly incorporated into this system. New MTA-SDK was integrated into one in cooperation with researchers from Nihon University and Chiba Institute of Technology. In the future, we will develop simulations under the new MTA-SDK and develop a development-execution environment.

8.5.2 *Distribution of Development-Execution Environment*

Until now, our technical results were published on the Internet, and source code, etc. (files compressed from directories containing necessary source code, etc.) were also distributed. However, simulation development-execution environment Simcast and information system development-execution environment MTA-SDK have been developed separately, and source code etc. have been published and distributed separately on the Internet. It is a hassle twice. In addition, in order to use them, it was necessary to download distribution files and compile them on Linux, so users were limited to engineering students and researchers.

However, as shown in the previous paragraph, since two development-execution environments were integrated into one new MTA-SDK this fiscal year, downloading is only once. In addition, the MTA-SDK will be published and distributed on the Internet as a "virtual machine" file instead of the source code.

https://www.theoreticalapproach.net/

The virtual machine file MTA-SDK.ova is distributed in a format common to virtualization software manufacturers, the OVF 1.0 standard. As a result, it is now possible to install commercial or free virtualization software "any OS," and then import this virtual machine to run the new MTA-SDK. Therefore, not only Linux users but also Windows users can use the development-execution environment, and the number of users can be expected to increase. It is also expected that the educational effect will be very high.

8.6 Publication of Practical Examples

Regarding the publication of practical examples, the following simulation development examples are introduced along with a rough theoretical explanation in Chapter 9 of Takahara et al. (2016).

8.6.1 *Price Adjustment System*

It is a simulation of a system (price adjustment system) for equilibrium price in classical microeconomics. Producers (sectors) and consumers (sectors) are modeled in the language CAST, each as a system of demand functions and supply functions. When executed, it behaves like a so-called spider's web. As a whole of composite systems, it is an example of a sequential processing system with feedback.

The introduced one defines producers and consumers as functional systems due to a space constraint. In actual research, we also simulate each as a problem-solving system (not a function). As a result of solving the problem for the price given by the adjuster of the meaning of Walras, it is a system to find the optimum production volume and optimum demand dose. In this way, an automaton containing a problem-solving system can be created (Asahi, 2009). Problem-solving is one of the features of the present stage in the simulation by the model theory approach, but it is not made use of in the publication stage.

8.6.2 Two-Body Dynamical System

It is a simulation of the motion of two bodies attracted by gravity in space. Gravity is the interaction between two bodies. The movement of each object can be specified as a state system. The gravitational force determines the acceleration, the acceleration determines the velocity, and the velocity determines the position.

Such models can be defined in the language CAST. As a whole of complex systems, it is an example of a concurrent processing system. Therefore, the composite system can be represented as shown in Fig. 8.2, and therefore modeled as shown in Eq. (8.2). When executed, the elliptical orbit of two objects can be obtained (drawn in the figure).

8.7 Conclusion

In this chapter, the theory and technology on simulation development by model theory approach were examined on the significance of the research results so far. In order to take advantage of the characteristics of the model theory approach described in Sect. 8.2, we improved language CAST so that there is a consistency between theory and practice while enriching the background theory (Sect. 8.3), and realized the speedup of simulation execution (Sect. 8.4). In addition, multiple development-execution environments were integrated into a new MTA-SDK, and it is published on the Internet as a highly versatile virtual machine (Sect. 8.5).

As a result, by implementing technological improvements while enriching the theory, it has become possible to take some consistency between theory and practice. It is the significance of this study that this achievement was succeeded. As we have seen in Sects. 8.3 and 8.4, the same structure of mathematical models and user models (computational models by language CAST) is a substantial technical basis.

Acknowledgments This content was originally published in Japanese language (Asahi, 2017). And it is translated into English by Asahi and partly revised for this book in September 2021. The author thanks "Keiei Ronshuu (Journal of Business Administration)" of Toyo University for granting permission for translation.

References

Asahi, T. (2009). Simcast: Model theory approach for simulation. *Keiei Ronshu (Journal of Business Administration, Toyo University), 73*, 33–51. Permalink: id.nii.ac.jp/1060/00004563/

Asahi, T. (2011). Model theory approach for simulation: Development procedure of autonomous distributed systems. *Keiei Ronshu, 78*, 177–188. Permalink: id.nii.ac.jp/1060/00004454

Asahi, T. (2012). Simulation model of interaction in two-body dynamical systems. *Keiei Ronshu, 80*, 31–38. Permalink: id.nii.ac.jp/1060/00004502/

Asahi, T. (2013). Formal models of connected systems in model theory approach. *Keiei Ronshu, 82*, 101–112. Permalink: id.nii.ac.jp/1060/00006348/

Asahi, T. (2014). Definition of defList in model description language CAST. *Keiei Ronshu, 84*, 41–50. Permalink: id.nii.ac.jp/1060/00006878/

Asahi, T. (2015). Models of many-component systems in model description language CAST. *Keiei Ronshu, 86*, 59–73. Permalink: id.nii.ac.jp/1060/00007949/

Asahi, T. (2017). Model theory approach for simulation: Improvements of model description language and integration of development environments. *Keiei Ronshu, 89*, 35–44. Permalink: id.nii.ac.jp/1060/00008573/

Asahi, T., Takahara, Y., Nakano, B., Saito, T., Shiba, N., & Chikuda, N. (2008). Model description language CAST for MIS development. *Journal of Japan Society for Management Information, 16*(4), 19–30. Retrieved from www.jasmin.jp/activity/books/ichiran.html. (in Japanese).

Takahara, Y., Asahi, T., & Hu, J. (2004). Application of MGST design approach to data mining systems: Case of I-0-0 type problem. *Journal of Japan Society for Management Information, 12*(4), 35–50.

Takahara, Y., Banerji, R., Asahi, T., & Shiba, N. (2008). Theoretical foundation for browser-based management information system development. *International Journal of General Systems, 27*(3), 275–304. (in English).

Takahara, Y., Saito, T., Asahi, T., & Shiba, N. (2007). *Formal method - model theory approach: Basis of information system development*. Nikka Giren Shuppan.

Takahara, Y., Saito, T., Asahi, T., Shiba, N., Chikuda, N., & Takagi, T. (2016). *Formal method - model theory approach: Basis of information system development* (Vol. 2). Nikka Giren Shuppan. [volume of practice].

Chapter 9
Declarative Modeling for Multimodal Processes Driven Distribution Networks

Zbigniew A. Banaszak

Abstract The presented issue concerns a declarative modeling based methodology aimed at assessment of possible mesh-like distribution networks carrying out multimodal processes flow. In order to achieve it, the grid network topology concept is used to model, analyze, and design of supply distribution networks incorporating multimodal processes paradigm. So as to guarantee congestion-free flow of multimodal processes encompassing the movement of items following delivery routes the deadlock-freeness for supporting the grid network consisting different modes of local transportation processes has to be guaranteed. Adopted declarative model of a mesh-like network describing forming it local and multimodal processes provides a framework enabling to determine the conditions sufficient to guarantee their cyclicity (i.e., blockage-free course). In that context, the main problem boils down to setting the sufficient conditions guaranteeing assumed processes performance while matching topological constraints of the network structure and following them constraints resulting from different speeds of supporting the local processes. The results obtained are illustrated in the results of case studies concerning AGVS and milk-run systems operation in different, mesh-like layouts of distribution networks including grid and fractal topologies as well.

Keywords Multimodal process · Mesh-like structure · Concurrent cyclic processes · Declarative modeling

9.1 Introduction

The proposal of a declarative modeling based methodology aimed at assessment of possible mesh-like distribution networks carrying out multimodal processes flow is the main objective of this chapter. In order to achieve it, the mesh network topology

Z. A. Banaszak (✉)
Faculty of Electronics and Computer Science, Koszalin University of Technology, Koszalin, Poland
e-mail: Zbigniew.Banaszak@tu.koszalin.pl

K. Kijima et al. (eds.), *Systems Research I*, Translational Systems Sciences 26,
https://doi.org/10.1007/978-981-16-9937-5_9

concept is used to model, analyze, and design of supply distribution networks incorporating multimodal processes paradigm. Examples of such processes can be found in different application domains such as production flows, intercity fright transportation supply chains, multimodal passenger transport network combining several unimodal networks (bus, tram, metro, train, etc.) as well as service domains (including passenger/cargo transportation systems, e.g., Automated Guided Vehicle Systems (AGVS), ferry, ship, airline, train networks, and also the data and supply media flows, e.g., cloud computing, oil pipeline and overhead power line networks) (Banaszak, 1997; Bocewicz & Banaszak, 2015; Bocewicz et al., 2020; Crama et al., 2000; Dawande et al., 2005; Kats & Levner, 1998; Li et al., 2016; Wallace & Yorke-Smit, 2020). A characteristic feature of multimodal processes (Bocewicz et al., 2012, 2013, 2014a; Bocewicz & Banaszak, 2015) is that their flows (e.g., following delivery routes) comprise fragments of local flows supported by different means of modalities (e.g., executed by separate, local delivery/handling processes). As a consequence, items moving along multimodal processes are moved by local unimodal means of distribution along sections of local process routes. In the general case, multilayered structures in which multimodal processes of the lower layers are used as local processes by multimodal processes of the higher layers based on them can be considered as well.

The smooth (undisturbed) flow of multimodal processes is conditioned by the deadlock-free implementation of the local unimodal processes supporting them. The blockages of the concurrently flowing local processes are followed by the conflicts occurring in situations caused by their simultaneous access to shared system resources, e.g., bus and train stations, transportation hubs, workstations, and warehouses. Consequently, in order to avoid the occurrence of blockages it is necessary to introduce appropriate mechanisms, e.g., employing dispatching rules, that synchronize the processes while guaranteeing a cyclic steady-state behavior of concurrently executed local processes. In that context, since cyclic flow is a deadlock-free one, the NP-hard problem of deadlocks handling may be treated as equivalent to the problem of cyclically executed local processes synchronization. It is worth noting that since the transportation processes executed along unimodal networks are usually cyclic, hence the multimodal processes supported by them have also periodic character. That means, the periodicity of multimodal processes flow depends on periodicity of unimodal (local) processes executed. In other words, the delivery period in a network formed by a set of local carriers depends on the cycle of this network. Consequently, the throughput of a delivery network is maximized by minimization of its cycle time.

It is assumed that the structure of the distribution network under consideration adopts topological assumptions following from numerous reports concerning commonly found regular networks, e.g., mesh-like or grid-like as well as fractal-like structures widely observed in different application domains (such as manufacturing, intercity freight transportation supply chains, passenger urban transport). Such topologies have been considered for over 20 years and are motivated by the expected advantages gained from their layout flexibility being vital to improve manufacturing (public transport) flexibility facing with expectations of shorter lead times, variable product (passengers, goods) mix and volumes (Barakou et al., 2015; Bocewicz et al., 2016, 2017a, b, c, 2019b; Montreuil, 1999; Venkatadri et al., 1997).

Abovementioned process classes and synchronization problems are represented here in declarative modeling framework which is based on the semantics of the natural systems being modeled rather than the algorithms that calculate their changing states. Therefore, declaratively expressed models are independent on architectural and software details, however only depend on the conceptualization (paradigm) adopted, and are thus easier to be exchanged and communicated as long as the basic conceptualization is agreed upon. Therefore, assuming that a mesh-like structure of a given distribution network limiting the course of local cyclical processes carried out in it, the main problem boils down to setting the conditions guaranteeing the local processes are deadlock-free as well as supported by them flow of multimodal processes. Searching for such conditions matching the topological constraints of the network structure and constraints resulting from different speeds of the local processes carried out in them, is aimed at assumed performance measures such as, in case of manufacturing systems, a production flow takt time, a flow time, a flow rate, and a job shop productivity.

To summarize, the new contributions provided to the currently available literature are (1) presenting a declarative modeling based methodology aimed at assessment of mesh-like distribution networks carrying out multimodal processes flow, (2) elaborating of a declarative reference model for mesh-like distribution networks and sufficient conditions guaranteeing congestion-free flow of delivery processes, and (3) developing of methods that allow planning multimodal deliveries in the distribution networks with grid or fractal topological structures.

The remaining of this chapter is organized as follows. Section 9.2 elaborates related works concerning modeling and control of concurrently flowing discrete processes focusing on both cyclic and multimodal processes. Section 9.3 presents the declarative approach to performance modeling of the mesh-like distribution networks with a particular focus on periodicity conditions. Section 9.4 provides the final conclusions followed by the description of future research.

9.2 Concurrent Discrete Processes

Process occurring in various fields of manufacturing application such as product flows in factories, city traffic flow, data flow in computer networks, etc., have one feature in common—they need to be executed concurrently. This means that concurrency is a common feature in the systems, where several processes can execute different actions simultaneously however often overlapping in time. Therefore, in case when the routes of individual processes go through common (shared) system resources, the blockade and/or congestion-free flow of interacting processes depends on assumed dispatching rules controlling (synchronizing) their access to such resources.

9.2.1 Systems of Concurrently Flowing Cyclic Processes

The considered class of concurrently realized discrete cyclic processes is observed in production planning, telecommunication transmission, and timetables determining. A representative example of the considered hereinafter class of Systems of Concurrently executed Cyclic Processes (SCCP) (Banaszak, 1997; Bocewicz & Banaszak, 2013; Majdzik & Obuchowicz, 2006; Polak et al., 2003, 2004) is the railway system. Single cyclic processes correspond to particular trains circulating in the connection network, stopping according to a specified timetable, on the stations spaced along the route traced in the railway network.

The timetable determining the arrival/departure time of the trains stopping at a given station is a solution to a cyclic scheduling problem, being so far a subject of many studies (Bocewicz & Banaszak, 2017; Bocewicz et al., 2017c, 2019a; Levner et al., 2010).

For the purpose of its illustration, consider the system shown in Fig. 9.1a and its graph model from Fig. 9.1b. A solution that represents a workstations' delivery services order which minimizes the total production cycle is sought.

9.2.1.1 Deadlock-Freeness

A model with Fig. 9.1b will be used to find a solution. To explain the further terminology of the process classes projected, the following illustrations of sequential

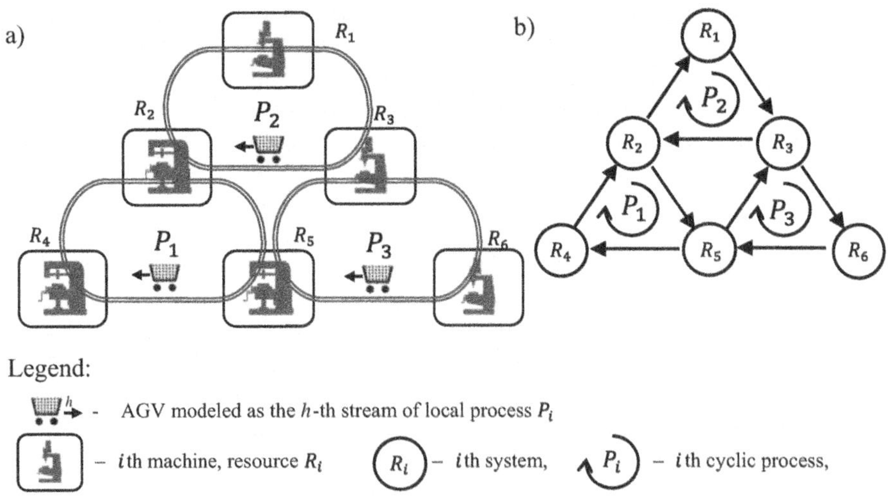

Fig. 9.1 Exemplary layout of flexible manufacturing system (**a**), and its graph model representation (**b**)

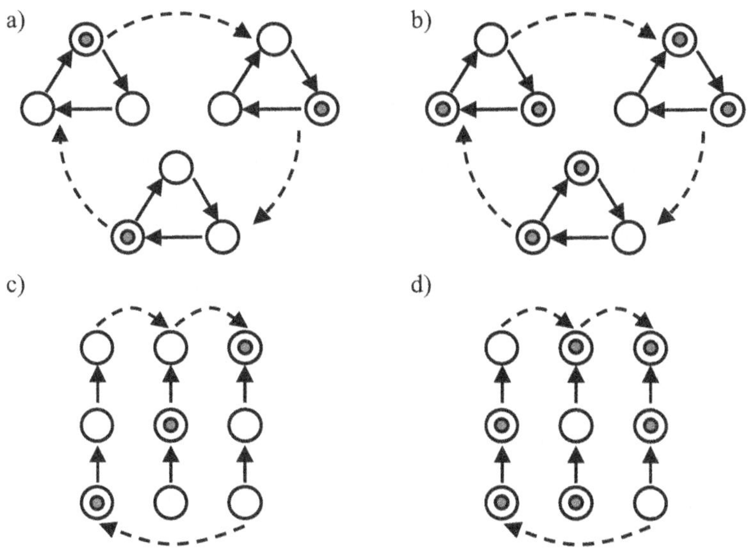

Fig. 9.2 Alternative representations of sequential (**a**), (**c**), and pipeline-like (**b**), (**d**) cyclic processes

Fig. 9.2a and serial or pipeline-like cyclic processes Fig. 9.2b that make up the SCCP, e.g., see Fig. 9.1b, are considered.

It is easy to see that the processes shown in Fig. 9.2a, b are formed by closed loops of transitions and are equivalent to their linear representations presented on Fig. 9.2c, d. What makes these representations different in the case of cyclical versions, it boils down to graphically emphasize the closed loop according to which the flow of the process is carried out. In turn, in the case of a linear representation, it boils down to a condition according to which the completion of the last operation of the process forces its first operation to start again. Noting this equivalence in further considerations the term cyclical process will cover both types of processes.

For deriving the cyclic conditions of SCCP class systems, let's consider the structure of an example instance of it shown in Fig. 9.3. Assumed allocation of processes carried out in a given instance of SCCP is represented by the sequence $(\mathrm{PA}_i(R_k), \ldots, \mathrm{PA}_j(R_1), \ldots, \mathrm{PA}_w(R_d))$ describing the current resource allocation, where:

$$\mathrm{PA}_i\,(R_k) = \begin{bmatrix} P_i \text{ if the } i\text{th process is allocated to the } k\text{th resource} \\ \Delta \text{ if no process is allocated to the } k\text{th resource} \end{bmatrix}$$

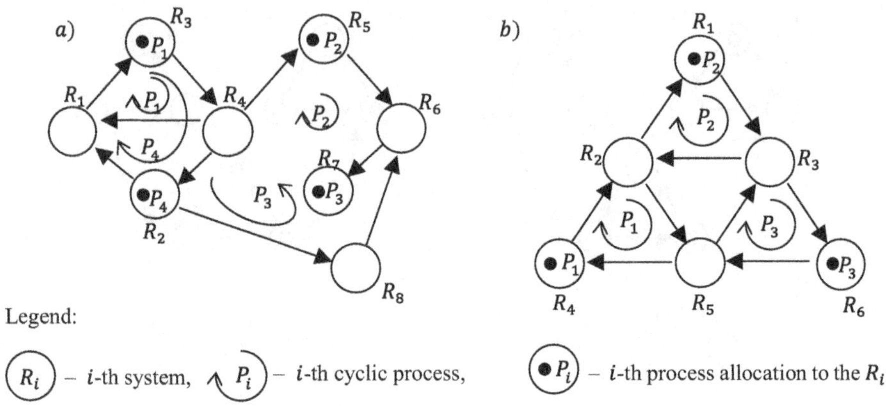

Fig. 9.3 Exemplary system belonging to the SCCP: illustrating a general case (**a**), and the following Fig. 9.1a layout (**b**)

Table 9.1 The sequence of processes allocation reached from the initial allocation PA_0 following priority dispatching rules: $\sigma_2 = (P_1, P_2)$, $\sigma_3 = (P_2, P_3)$, $\sigma_5 = (P_3, P_1)$

	$R_1, R_2, R_3, R_4, R_5, R_6$
PA_0	$(P_2, \Delta, \Delta, P_1, \Delta, P_3)$
PA_1	$(\Delta, P_1, P_2, \Delta, P_3, \Delta)$
PA_2	$(\Delta, P_1^*, P_2^*, \Delta, P_3^*, \Delta)$

P_i* means that the *i*th process occupying the relevant resource is awaiting to release another (in the order given by its route), P_i means that the *i*th process is executed on the resource, Δ means that no process is executed nor awaiting on the resource

Example of a processes allocation corresponding to the one shown in Fig. 9.3b illustrates the following sequence PA_0, where the symbol Δ indicates no resource assignment.

$$R_1, R_2, R_3, R_4, R_5, R_6$$

$$PA_0 = (P_2, \Delta, \Delta, P_1, \Delta, P_3)$$

Assuming that $PA_0 = (P_2, \Delta, \Delta, P_1, \Delta, P_3)$ sequence is the initial allocation of processes in the system under consideration, it can be shown that it leads to a deadlock allocation, in which the process p_1 requests access to the resource R_3, the process P_2 requests access to the resource R_1, and the process P_3 requests access to the resource R_2. The closed loop of mutual requests is illustrated in Table 9.1.

The considered case of processes blocking results from the adopted method of resource conflicts resolution due to which the order of processes allocation on each the *i*th resource is controlled by the priority dispatching rule $\sigma_i = (P_j, P_a,$

$P_b, \ldots, P_k, \ldots, P_n$) determining the order in which after execution of an operation performed by process P_j on the resource R_i, the next access to R_i is allocated to P_a, then to P_b and so on. Note that such synchronization rules (protocols) can be treated as processes coordinating (mapping the prescribed dispatching priority rules) the access of processes competing for access to shared resources. Therefore, the set of Priority Dispatching Rules (PDRs) $\sigma_2 = (P_1, P_2)$, $\sigma_3 = (P_2, P_3)$, $\sigma_5 = (P_3, P_1)$ results in the behavior described by the sequence of processes allocation collected in Table 9.1.

In general, however in a given structure of cyclic processes PDRs may result in different ways of processes flow leading or not to its blockade. Of course, the deterministic nature of the rules results in deterministic behavior of SCCP. That means that the connectivity of SCCP structures implies the only one type of behavior corresponding to a given set of PDRs which is the processes blockage or cyclically repeating sequence of processes allocation. Therefore the system cycle (its cyclicality) is determined by the choice of priority rules used in it, and in general the adoption of different dispatching rules may lead to periodic behaviors characterized by different repeatability period T. In the considered case, see Fig. 9.3b for the initial allocation $PA_0 = (P_2, \Delta, \Delta, P_1, \Delta, P_3)$ and the following set of PDRs $\{\sigma_2 = (P_1, P_2), \sigma_3 = (P_2, P_3), \sigma_5 = (P_1, P_3)\}$ the cyclic steady state of processes allocations containing repeating sequence of allocations PA_2–PA_7 is followed by transient period consisting two allocations PA_0 and PA_1, respectively. Because processes allocations PA_2–PA_{10} repeat every nine allocations therefore, the system cycle thus calculated is equal to $T = 9$. The described course illustrates sequence of processes allocations collected in Table 9.2.

To sum up the study of the relationships linking the behavior of the SCCP system with its initial processes allocation, the structure and the set of dispatching rules boils down to setting conditions (synchronization mechanisms) enabling the synthesis of distributed control procedures for SCCP.

Table 9.2 The sequence of processes allocation reached from the initial allocation PA_0 following priority dispatching rules: $\sigma_2 = (P_1, P_2)$, $\sigma_3 = (P_1, P_3)$, $\sigma_5 = (P_2, P_3)$

	$R_1, R_2, R_3, R_4, R_5, R_6$
PA_0	$(P_2, \Delta, \Delta, P_1, \Delta, P_3)$
PA_1	$(\Delta, P_1, P_2, \Delta, \Delta, P_3)$
PA_2	$(\Delta, \Delta, P_2, \Delta, P_1, P_3)$
PA_3	$(\Delta, P_2, \Delta, P_1, \Delta, P_3)$
PA_4	$(P_2, \Delta, \Delta, P_1, P_3, \Delta)$
PA_5	$(P_2, P_1, P_3, \Delta, \Delta, \Delta)$
PA_6	$(P_2, \Delta, \Delta, \Delta, P_1, P_3)$
PA_7	$(\Delta, \Delta, P_2, P_1, \Delta, P_3)$
PA_8	$(\Delta, P_2, \Delta, P_1, P_3, \Delta)$
PA_9	$(P_2, \Delta, P_3, P_1, \Delta, \Delta)$
PA_{10}	$(P_2, P_1, \Delta, \Delta, \Delta, P_3)$

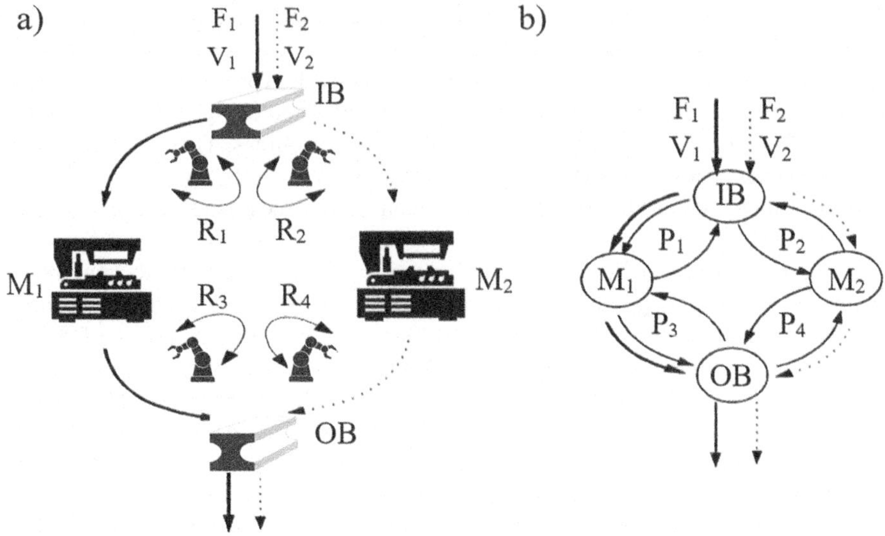

Fig. 9.4 FMM: (**a**) a layout scheme; (**b**) a processes interaction model. R_i—the ith robot; M_j—the jth machine tool; IB (OB)—the input (output) buffer; P_i—the ith repetitive process of workpiece handling and transportation; F_i—the ith production route; V_i-the lot production associated with the ith production route

9.2.1.2 Mesh-Like Network Structures

To enter the area of mesh-like class structures let's consider a Flexible Manufacturing Module (FMM) as shown in Fig. 9.4a composed of feeding robots R_1 and R_3 that serve to pick up the workpieces from the input buffer IB, and then to put them into machine tool M_1 or M_2 in predefined positions. The collecting robots R_2 and R_4 serve to pick up the machined products from a machine tool and to put them into the output buffer OB. Two kinds of products are processed along the production routes F_1 and F_2, respectively. IB, OB, M_1, and M_2 are common resources, and the co-operating robots compete for access to them. The direction of a cyclic process is shown by thin solid arrows; the direction of the ith production route followed by corresponding multimodal process is shown by dotted arrows and thick solid arrows (Banaszak, 1997).

The protocols managing processes to access FMM common resources while guaranteeing its cyclic steady-state behavior (i.e., resulting in the repetitive character of the material flow), have to follow the assumptions:

- Each elementary component process is treated as a repetitive one that performs a non-empty sequence of pickup and place operations.
- Only one operation can be executed at a time on each system resource.

- Coupled processes that compete for access to the system's shared (non-preemptive) resources can be either executed or suspended on one of the associated resources.
- A process can be suspended on a given resource only if the next resource to be used is busy.
- A suspended process can be executed only if the next resource to be used is not busy.

In the mesh-like structure of the flexible manufacturing system (FMS), see Fig. 9.5a, composed of FMMs (see Fig. 9.4a), the component parts interact with one another via shared resources (e.g., M_1, M_2, B_1 and B_3). Interacting sequential repetitive processes compete for access to shared resources using a mutual-exclusion protocol. A similar case of process competition occurring in this case concerns multimodal processes, see Fig. 9.5b. Two multimodal processes corresponding to production routes marked by dashed arrows (following sequence B_1, M_2, B_4, M_6, B_5) and solid arrows (following sequence B_2, M_2, B_3, M_4, B_5) compete with access to common shared M_2. Designation of a set of protocols (i.e., PDRs) controlling access to shared resources so as to guarantee a desired performance of a whole system (i.e., the qualitative guaranteeing a starvation-free and deadlock-free system run, and quantitative, i.e., satisfying desired values) corresponds to a distributed control design.

Similar to the above-mentioned situations related to the need for synchronization (i.e., distributed control) of concurrently executed periodic processes are typical for a class of Periodic Vehicle Routing Problems (PVRP). PVRP is an NP-hard problem in which delivery routes are constructed over a period of time (Kamoun & Sriskandarajah, 1993; Kats & Levner, 1998; Levner et al., 2010; Li et al., 2016). Its objective is to minimize travel costs or the total travel distance required to visit all

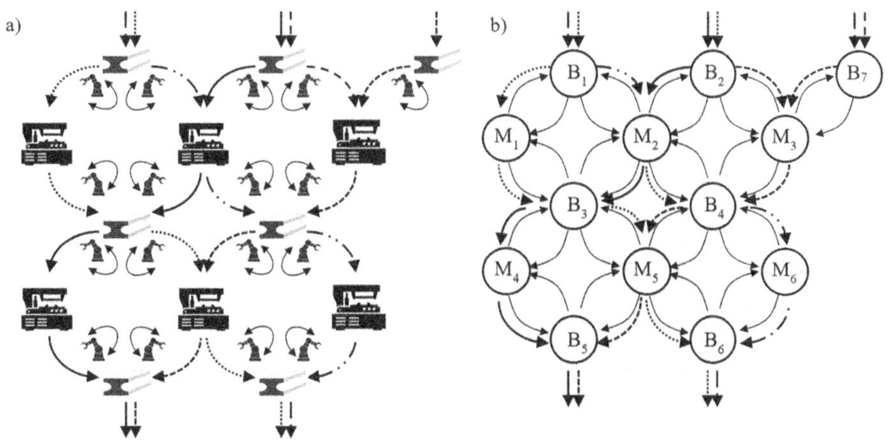

Fig. 9.5 The mesh-like structure of FMS: (a) a layout scheme; (b) a processes interaction model

customers while following the assumed schedule of frequency visits within a given time horizon.

A particular role in this class of problems is played by structure of the transport systems assumed in them. Especially structures with a regular, recursive morphology typical of tree or mesh (grid) topologies, which include urban transport systems, are the subject of intensive research (Bocewicz et al., 2015, 2017c; Montreuil, 1999; Venkatadri et al., 1997). The main advantages following from the regular structure of supply network layout are flexibility and robustness which are vital to the improvement of robustness of supply/distribution networks. Moreover, the development of urban agglomerations, and in particular the morphology of urban regions, is subject to the laws of recursion, which are best modeled by fractal structures. The consequences of this fact can be used both in predicting the needs related to the expansion of the existing transport infrastructure, and in planning new industrial and/or urban agglomerations (Bocewicz et al., 2017c; Wójcik et al., 2016).

To illustrate the concepts of the mesh-like structures discussed hereinafter, let's consider an example of grid-like structures composed of clusters (involved in a fractal growth) which are identical repeating substructures, called Elementary Covering Structures (ECS) (Bocewicz et al., 2016). The proper example provides ECS from Fig. 9.6b, where local processes indicate the direction of flow of local traffic (transport modes). The adopted ECS standard can be used to build more complex configurations such as grid and fractal structures following: chaotic (a), see Fig. 9.6a, and fractal (b), see Fig. 9.6b patterns.

Regular structures supporting mass customized services are found in different application domains (intercity fright transportation supply chains, multimodal passenger transport network combining several unimodal networks (bus, tram, metro, train, etc.) as well as data and supply media flows, e.g., cloud computing, oil pipeline, and overhead power line networks) (Banaszak & Krogh, 1990; Bocewicz et al., 2016, 2017b, 2019b). An example of city districts growth and its corresponding fractal representation provides another example of transportation network growth, see Fig. 9.7a (Bocewicz et al., 2016).

The transportation network obtained after seventh iterations can be seen as a network of bus/trams lines providing periodic service along cyclic routes. In that context, the problems arising in such transportation networks concern of routing and scheduling of multimodal processes of passengers flows which are in general case also NP-hard problems. It should be noted however, the passenger travel schedules can be estimated easily while taking into account cyclic behavior of both: local transportation modes and the whole transportation network. That is because, since the transportation processes executed by particular lines are usually cyclic, hence the multimodal processes supported by them have also periodic character. That means, the period of multimodal passenger flow depends on a period of the relevant transportation network, especially on ECSs.

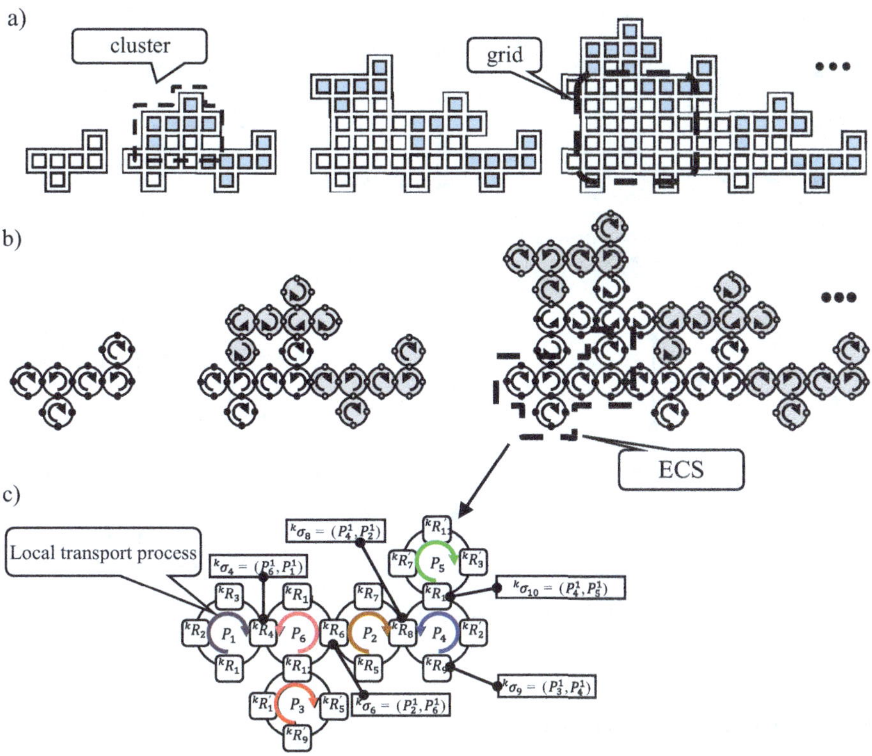

Fig. 9.6 Grid-like structure (**a**) composed of ECSs (**b**) created from local processes (transport modes). (**c**) ${}^{k}R_r$—rth resource in the kth ECS, P_i—local transportation process of the ith transport mode, ${}^{k}\sigma_r$—dispatching priority rule assigned to ${}^{k}R_r$

9.2.2 *Multimodal Processes*

As already mentioned, the course of multimodal processes in particular, their cycle and the delivery takt depend on the structure and the cycle time of the supporting them SCCP. The topological structure of considered SCCP system can be a mosaic of different grid structures synchronized by properly chosen dispatching rules (Banaszak & Polak, 2005; Bocewicz et al., 2009; Polak et al., 2003, 2004). However, such rules may be set up when the cycles of the subsystems corresponding to the differing fragments of mosaic the following condition holds (Bocewicz et al., 2017b, 2018, 2019b; Bocewicz & Banaszak, 2017): $T = \mathrm{NWW}(T_1, T_2, \ldots, T_i, \ldots, T_n)$, where T is the period of whole SCCP, T_i is the period of the ith subsystem of SCCP.

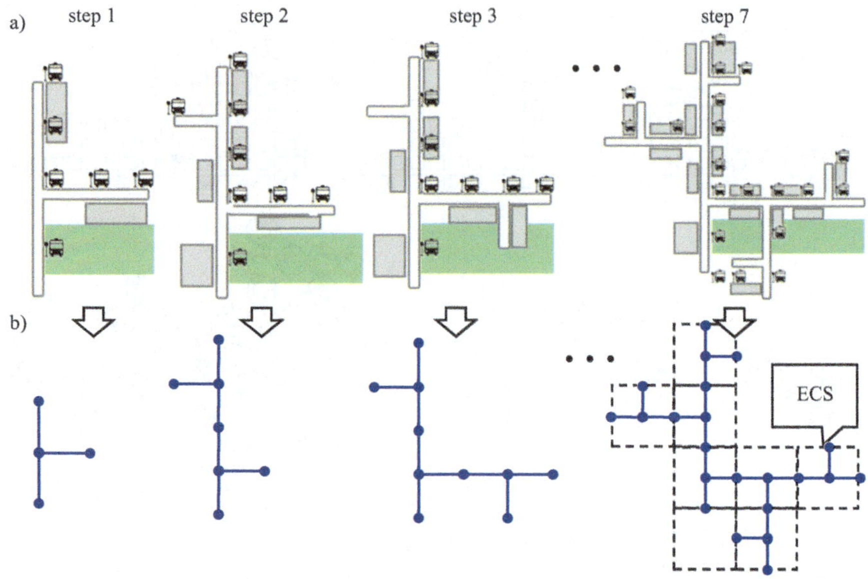

Fig. 9.7 Exemplary snapshots of city districts growth (**a**) and corresponding to the fractal model representations (**b**)

9.2.2.1 Distribution Networks

Consider SCCP being of the seventeenth iteration of the fractal pattern, from Fig. 9.7b shown in Fig. 9.8 which can be seen as composed of "+" shaped ECS, see Fig. 9.9a. Each edge of ECS is replaced by a cyclic process executed along four resources. Two resources (distinguished by dots) correspond to vertices and model interchange stations (hubs) shared by other cyclically operating lines. The remaining two, so-called unshared, resources correspond to pavements (tracks) traveled by considered transportation mode, see Fig. 9.9b.

Apart from local processes, two multimodal processes (i.e., processes executed along the routes consisting parts of the routes of local processes): mP_1, mP_2 are considered. The transportation route depicted by the red color line corresponds to the multimodal process mP_1 supported by different transportation modes (busses, trams, metro, etc.), which in turn encompass local transportation streams P_2 and P_4. This means that the transportation routes specifying how a multimodal process is executed can be considered as composed of parts of the routes of local cyclic processes. To ensure the periodicity of multimodal processes implemented in different ECS, it must be demonstrated that there is such a way of their composition which guarantees cyclic behavior of entire SCCP.

As another example, consider a multi-item batch flow production system in which in-plant transport operations are organized in milk-run loops (Fig. 9.10). Logistics trains (tugger trains) traveling along fixed routes are used as in-plant

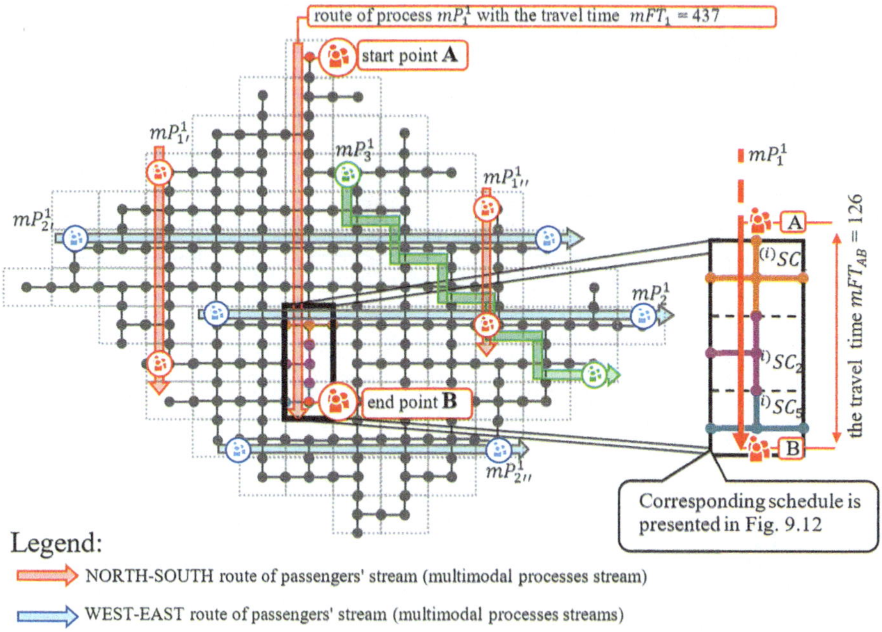

Fig. 9.8 SCCP being of the seventeenth iteration fractal pattern from Fig. 9.7b

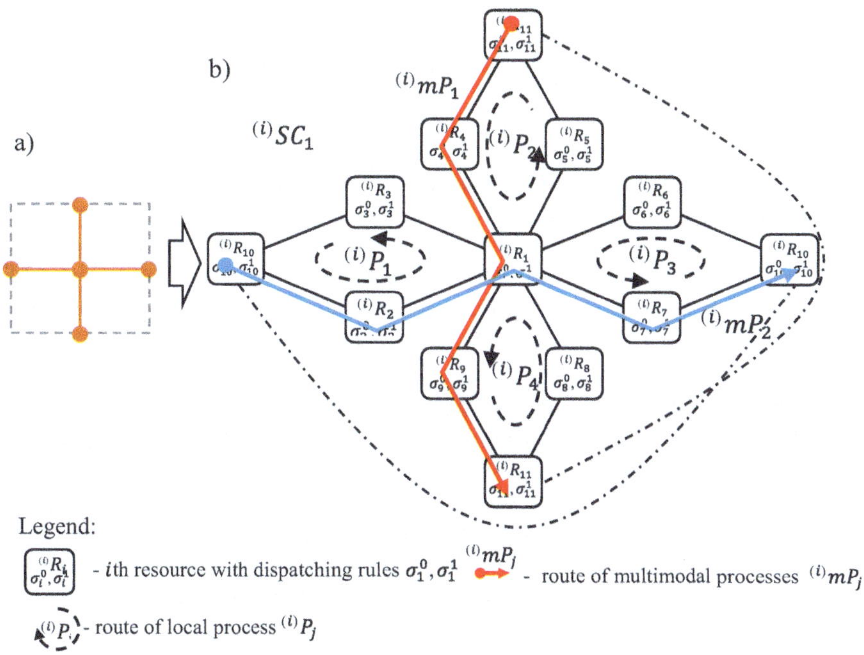

Fig. 9.9 ECS following "+" shape

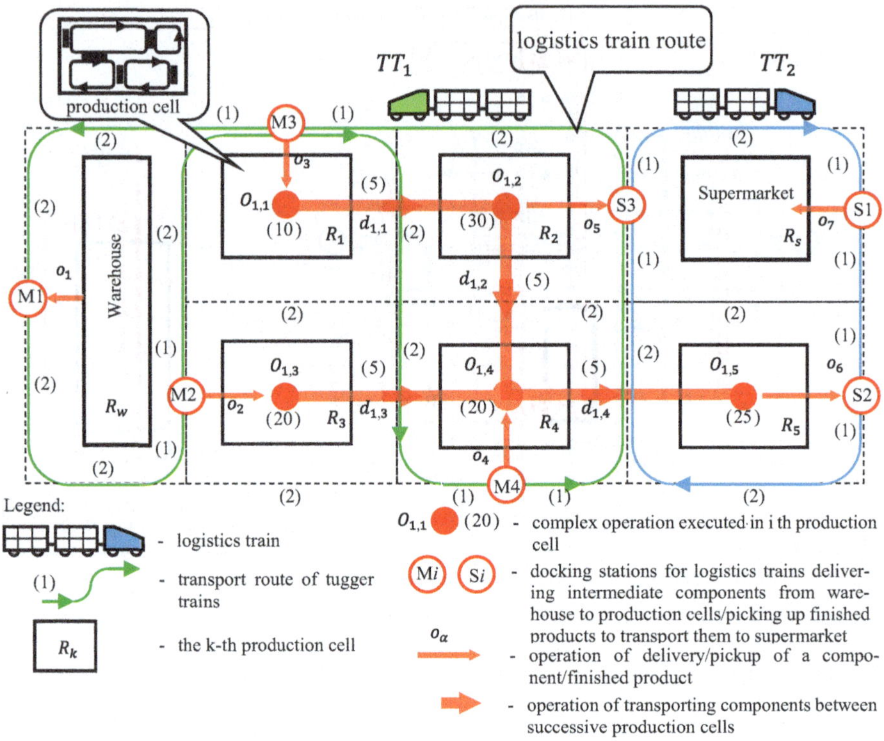

Fig. 9.10 Layout of the shop floor with marked production flow and milk-run routings

transport means to deliver the required quantity of materials to workstations within given time windows. It is worth noting that the work parts leaving the supermarket and then processed on subsequent workstations are part of various products stored in the warehouse. Inter-workstation transport operations related with the production of various products, i.e., creating different multimodal processes, are supported by different logistics trains moving along different closed loop routes. To be more precise a multimodal process is understood here as a workstation-to-workstation production flow process, whose sections are local, cyclically repeated, milk-run tours. The basic problem is to find a method by which to organize timely and congestion-free movement of the logistics trains used to deliver/pick up ordered goods (Bocewicz et al., 2019a).

In a system with the structure shown in Fig. 9.10, a way of distributing supplies of specific groups of goods in specified quantities, in the given time windows, to the given group of workstations is sought. In other words, we are faced with a question: Do there exist, in the given system, routes of logistics trains and the associated delivery schedules that guarantee timely delivery of the materials necessary for the production process to be completed?

9.2.2.2 Flow Schedules

Let's focus only on one case from the previous section, and related to multimodal process distinguished by red color line, see Figs. 9.8 and 9.9. In order to determine the time needed to travel on the road connecting the point A and B, the ECS cycle time should be calculated. In the case under consideration, it is 48 u.t. which results from the value of the adopted equal to 12 u.t. cycle of elementary process, see Fig. 9.11. Consequently, the takt time of multimodal process mP_1 is equal to 48 u.t., i.e., the stream of each multimodal process arrives to each resource only once on the one period of ${}^{(i)}m\alpha_1 = 48$.

The finally obtained schedule illustrating operation of transportation modes and a way they provide passengers' transportation route between points A and B is shown in Fig. 9.12. The travel time between points A and B the passengers have to spent while traveling along the route of multimodal process mP_1 – North-South is equal to 3·48–18 = 126 u.t., where: $c \cdot {}^{(i)}m\alpha_1 - \Delta F_1, {}^{(i)}m\alpha_1$—the period of multimodal processes executed by networks of regular structures, ΔF_k—the buffer time encompassing the difference between completion time of a travel and a beginning time of the subsequent one, c—the number of ECSs consisted by a transportation route. Therefore, the whole travel time along transportation route of mP_1 is equal to 437 u.t.

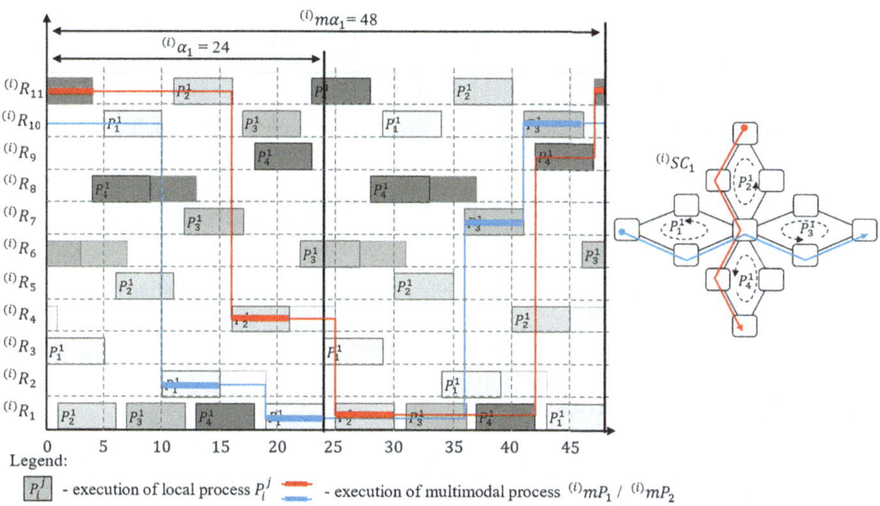

Fig. 9.11 The schedule ${}^{(i)}X'_1$ of SCCP following ${}^{(i)}SC_1$ from Fig. 9.9

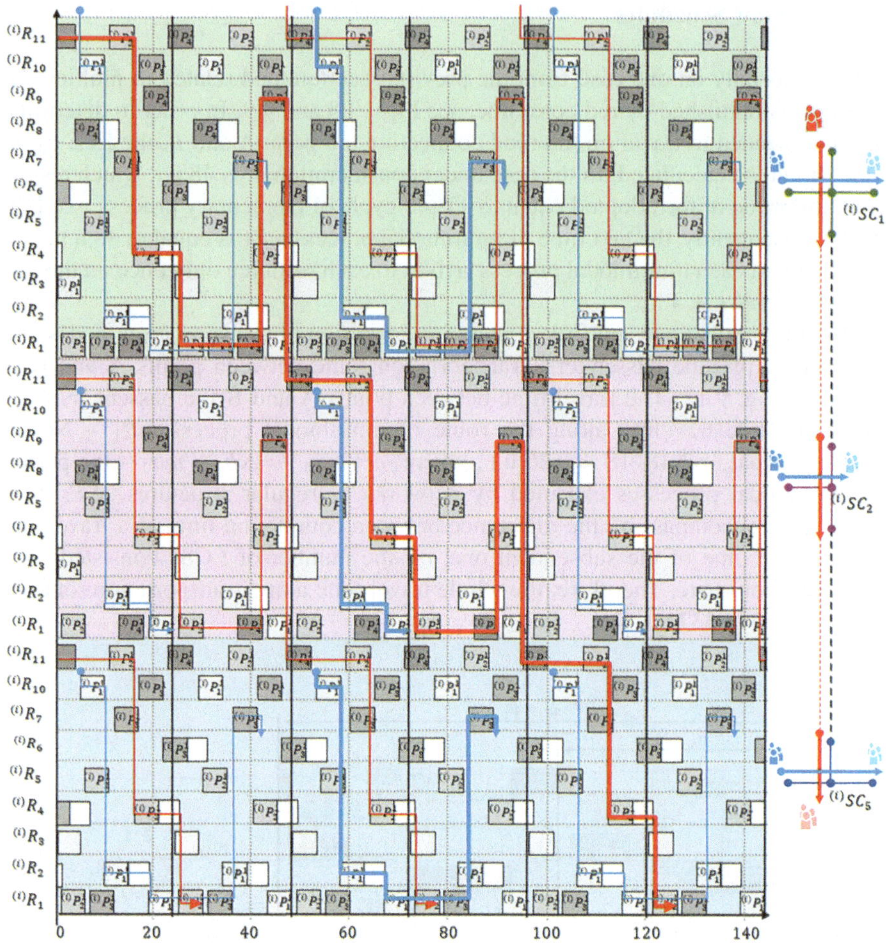

Fig. 9.12 The Gantt's chart of a part of fractal-like network from Fig. 9.8 including both local and multimodal processes

9.3 Performance Modeling

On the quality of the functioning evaluation of a given system the decisive influence has its model used in course of its assessment. Available model classes are divided into those using an imperative and those based on a declarative approach. Differences between both kind of models boils down to the fact that imperative one focus on *how* (to solve) while declarative focus on *what* (to describe/specify). Due to the nature of the research carried out, covering both the analysis and synthesis of SCCP issues a declarative modeling framework is implemented.

9.3.1 Declarative Modeling

A declarative modeling based constraint programming approach offers modeling features and solution methods that are unavailable in mathematical programming but are often flexible and efficient for scheduling and other combinatorial problems. The user just states the problem as a constraint satisfaction problem (CSP) and a generic solver finds a solution without additional programming.

9.3.1.1 Constraints Satisfaction Problem

In declarative models, the focus is on what the solution should be, i.e., instead of specifically describing how the searching process has to work, only the essential problem characteristics are described. In that context, the presented so far types of problems can be considered in terms of constraints satisfaction problems of the following form:

$$\mathrm{PS} = ((\mathrm{X}\,\mathcal{D}),\ \mathcal{C}), \tag{9.1}$$

where:

X—a set of decision variables,
$\mathcal{D}$—a finite set of decision variable domains,
$\mathcal{C}$—a set of constraints specifying the relationships between decision variables.

What is sought is a solution in which the values of all decision variables X satisfy all constraints $\mathcal{C}$.

Depending on the way the decision variables are declared, the reference problem (9.1) can be used to formulate analysis or synthesis problems in terms of a constraint satisfaction problem framework. The questions following problems of both categories can be stated as: Is it possible to make supplies which meet customer demands in a transport network with a preset structure? Is there a transport network structure that ensures deliveries which meet user expectations? To illustrate these variants, consider an example of the in-plant milk-run distribution system from Fig. 9.10 (Bocewicz et al., 2019a).

For the sake of further discussion, it is assumed that the problem whose aim is to assess the quality (e.g., just-in-time deliveries) of distribution processes, will be interpreted as an analysis problem, and the problem whose aim is to choose a variant of logistics train routes will be treated as a synthesis problem. In this context, the problem of analysis can be formulated as (9.2) (Bocewicz et al., 2019a):

$$PS_A = ((X, \mathcal{D}), CSR), \tag{9.2}$$

where:

$X = (X', Xs', \alpha')$—a cyclic schedule of process operations, where: $X' = (x_\lambda \mid \lambda = 1,\ldots,\omega)$, x_λ—starting time of o_λ operation delivering/loading/unloading of materials to/at workstations o_λ, $Xs' = (xs_\lambda \mid \lambda = 1,\ldots,\omega)$, xs_λ—time of release of a resource occupied by operation o_λ, α'—takt time of local process operations,

$\mathcal{D}$—a finite set of decision variable domains $\{X', Xs', \alpha'\}$, where: $\alpha', x_\lambda \in \{0,\ldots,T\}$, $xs_\lambda \in \{0,\ldots,2T\}$, and T stands for a planning horizon, i.e., the "time window" in which deliveries are made,

$\mathcal{CSR}$, a set of constraints specifying the relationships between operations of processes implemented in milk-run cycles and the processes executed along the technological routes of individual products. Particularly constraints for: operations of processes executed along the technological routes of individual products, local processes (transport operations of logistics trains), and processes executed in milk-run cycles and those executed along technological routes of individual products.

The resulting schedule X determines the admissible timetable of logistics trains guaranteeing timely delivery of materials to workstations when meeting the assumed constraints. This means that the admissible solution sought exists only under certain conditions determined by domains of decision variables and linking them constraints.

In turn the synthesis problem can be stated as follows (9.3) (Bocewicz et al., 2019a):

$$PS_{RE} = ((\{X, B, F\}, \mathcal{D}), \mathcal{CRE}), \tag{9.3}$$

where:

$\{X, B, F\}$—a set of decision variables including, where: X—a cyclic schedule of process operations executed in milk-run cycles, $X = (X', Xs', \alpha')$, where: $X' = (x_\lambda \mid \lambda = 1,\ldots,\omega)$, x_λ—starting time of o_λ operation delivering/loading/unloading of materials to/at workstations o_λ, $Xs' = (xs_\lambda \mid \lambda = 1,\ldots,\omega)$, xs_λ—time of release of a resource occupied by operation o_λ, α'—takt time of local process operations,

B, F—sequences specifying the order in which operations $o\lambda$ are executed by the successive processes run in milk-run loops, where:

$B = (b_1,\ldots,b_\lambda,\ldots,b_\omega)$—sequence of preceding operations, $b_\lambda \in \{0,\ldots,\omega\}$, b_λ is an index of the operation that precedes o_λ (operations ob_λ and o_λ are executed by the same process implemented in a milk-run cycle), $b_\lambda = 0$ means that o_λ is the first operation in the system's cycle,

$F = (f_1, \ldots, f_\lambda, \ldots, f_\omega)$—sequence of following operations, $f_\lambda \in \{1,\ldots,\omega\}$, f_λ—index of the operation that follows o_λ, (operations o_λ and *of* λ are executed by the same process implemented in a milk-run cycle. Each pair (B, F) is matched with exactly one pair from the set of routes and operations of processes executed in milk-run cycles.

$\mathcal{D}$—a finite set of decision variable domains $\{X, B, F\}$: $x_\lambda \in \{0,\ldots,T\}$, $xs_\lambda \in \{0,\ldots,2T\}$, $b_\lambda \in \{0,\ldots,\omega\}$, $f_\lambda \in \{1,\ldots,\omega\}$, where: ω—number of operations of processes executed in milk-run cycles,

$\mathcal{CRE}$—a set of constraints specifying the relationships between the processes implemented in milk-run cycles and the processes executed along the technological routes of individual products, particularly constraints for processes executed: along the technological routes of individual products, in milk-run cycles, and in milk-run cycles and processes executed along technological routes.

Sought are decision variables B, F determining parameters of delivery processes structure (routes of local processes) which are to guarantee the existence of schedule X that will enable timely delivery of materials to workstations (in accordance with the starting times of operations of processes executed along the technological routes of individual products). In other words, assuming that the set of processes executed in milk-run cycles P (by a fleet of logistics trains), process execution times t_λ, process flow times $d_{\lambda,\beta}$, and the starting times of the operations of processes executed along the technological routes of individual products $mx_{i,j}$ (dates of expected deliveries) are all known, to solve problem PS_{RE} (9.3), it is enough to determine such values (determined by domains $\mathcal{D}$) of decision variables B, F (routes of local processes) and X (schedule of local processes), for which all constraints given in the set $\mathcal{CRE}$ are satisfied.

Of course, the admissible solution sought exists only under certain conditions determined by domains of decision variables and linking them constraints.

Implementation of the above presented CSP-driven approach in an interactive DSS environment can be used for online prototyping of supply cycles in a milk-run system. In other words, depending on the choice of the operation mode (i.e., depending on the type of synthesis/analysis problem under consideration), such way allows one to quickly assess various variants of logistic trains routing.

9.3.1.2 States Space

Operations in cyclic processes are executed along sequences that repeat an indefinite number of times. Performance of such systems belonging to the SCCP class (i.e., forming the connected networks of processes sharing commonly used resources) depends on resource conflict resolution (processes synchronization) mechanism. The most commonly used mechanisms boils down to assignment of so-called dispatching priority rules (DPRs). As it was already mentioned, the behavior (functioning observed through sequences of states) of a SCCP depends on its structure and the initial processes allocation adopted in it.

To explain the concept of SCCP state let's assume that they are composed of the set of processes $P = \{P_1, P_2, \ldots, P_n\}$, where each process P_i executes periodically a sequence of operations using resources occurring along a given process route $p_i = \left(R_{j_1}, R_{j_2}, \ldots, R_{j_{\mathrm{lr}(i)}}\right)$, $j_k \in \{1, 2, \ldots, m\}$, where $\mathrm{lr}(i)$ denotes a length of cyclic process route and m denotes number of resources, and $R_{j_k} \in R$, where

$R = \{R_1, R_2, \ldots, R_m\}$. The time $t_{i,j}$ of operation executed on R_j along P_i, is defined in domain of uniform time units ($\mathbb{N}$—set of natural numbers, i.e., $t_{i,j} \in \mathbb{N}$). So, the sequence $T_i = \left(t_{i,j_1}, t_{i,j_2}, \ldots, t_{i,j_{l(i)}}\right)$ describes the operation times required by P_i. To each common shared resource $R_i \in R$ the DPR $\sigma_i = \left(P_{j_1}, P_{j_2}, \ldots, P_{j_{\mathrm{lp}(i)}}\right)$, $j_k \in \{1, 2, \ldots, n\}$, $P_{j_k} \in P$ is assigned, where $\mathrm{lp}(i) > 1$, $\mathrm{lp}(i)$ is a number of processes dispatched by σ_i. For the purpose of future considerations, it is assumed that the all operation times are equal to a unit operation time (noted as: u.t.), $\forall i \in \{1, \ldots, n\}$, $\forall j \in \{1, \ldots, \mathrm{lr}(i)\}$, $(\mathrm{crd}_j T_i = 1 \text{ u. t})$.

Let's define the kth state S^k as a triple (9.4) composed of the sequence of: processes allocations A^k, semaphores (encompassing the rights access to individual resources) Z^k, and semaphore indices Q^k:

$$S^k = \left(A^k, Z^k, Q^k\right), \tag{9.4}$$

where: $A^k = (a_1{}^k, a_2{}^k, \ldots, a_m{}^k)$ is the processes allocation in the kth state (m is a number of resources occurring in the SCCP). Each element $a_i{}^k$ means the process allotted to the ith resource R_i in the kth state: $a_i{}^k \in P \cup \{\Delta\}$, $P = \{P_1, P_2, \ldots, P_n\}$ is the set of processes, $a_i{}^k = P_g$ means, the ith resource R_i is occupied by the process P_g, and $a_i{}^k = \Delta$—the ith resource R_i is unoccupied. In the case considered (see Fig. 9.13a) the processes allocation is specified by the sequence: $A^0 = (\Delta, P_2, \Delta, P_1, \Delta, P_3)$.

$Z^k = (z_1{}^k, z_2{}^k, \ldots, z_m{}^k)$ is the sequence of semaphores corresponding to the kth state, where $z_i{}^k \in P$ means the name of the process (specified in the ith DPR σ_i, allocated to the ith resource) allowed to occupy the ith resource R_i. For instance $z_i{}^k = P_g$ means that at a moment the process P_g is allowed to occupy the ith resource R_i. For the SCCP from Fig. 9.13a the sequence of semaphores has the following form: $Z^0 = (P_2, P_2, P_2, P_1, P_3, P_3)$.

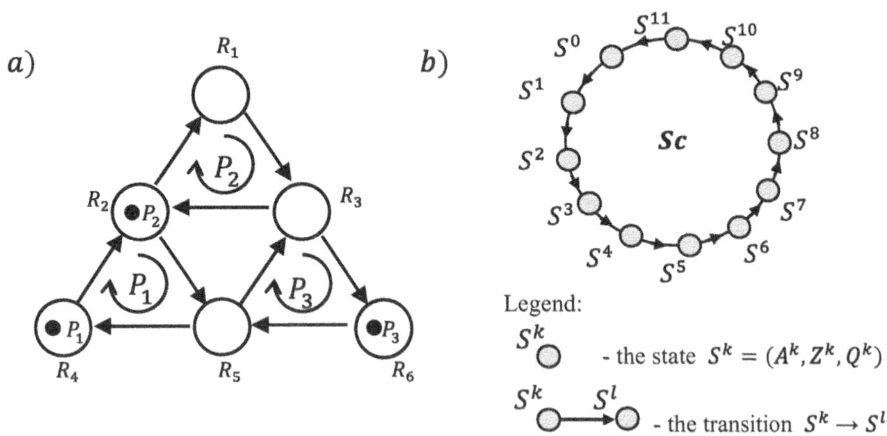

Fig. 9.13 The SCCP (**a**), and its cyclic steady-state space S_c (**b**) following $\sigma_2 = (P_2, P_1, P_2, P_1, P_1)$, $\sigma_3 = (P_2, P_3, P_2)$, $\sigma_5 = (P_3, P_1, P_1, P_1)$

$Q^k = (q_1{}^k, q_2{}^k, \ldots, q_m{}^k)$ is the sequence of semaphore indices corresponding to the kth state, where $q_i{}^k$ means the position of the semaphore $z_i{}^k$ in the DPR σ_i: $z_i{}^k = \text{crd}_{(q_i{}^k)}\sigma_i, q_i{}^k \in \mathbb{N}$ ($\text{crd}_i D = d_i$, for $D = (d_1, d_2, \ldots, d_i, \ldots, d_w)$). For instance $q_2{}^k = 2$ means the second position occupied by P_1 in the DPR σ_2, where: $P_1 = z_2{}^k$.

In the case of the SCCP from Fig. 9.13a for the adopted set of DPRs $\sigma_2 = (P_2, P_1, P_2, P_1\ P_1)$, $\sigma_3 = (P_2, P_3, P_2)$, $\sigma_5 = (P_3, P_1, P_1, P_1)$ and the initial state $S^0 = (A^0, Z^0, Q^0)$ such that $S^0 = ((\Delta, P_2, \Delta, P_1, \Delta, P_3), (P_2, P_2, P_2, P_1, P_3, P_3), (1, 1, 1, 1, 1, 1))$ the only directly reachable state $S^1 = ((P_2, \Delta, \Delta, P_1, P_3, \Delta), (P_2, P_1, P_2, P_1, P_3, P_3), (1, 2, 1, 1, 1, 1))$ there exists. The graphical illustration of the resultant cyclic steady-state space S_c generated by S^0 while following the transition $S^0 \rightarrow S^1$ is a transition state function $\delta : \mathbb{S} \rightarrow \mathbb{S}$ (Bocewicz & Banaszak, 2013), which is shown in Fig. 9.13b).

Assuming that a cyclic steady-state period T_c is defined in the following way: $T_c = \| S_c \|$ the searching for a cyclic steady state S_c in a given SCCP can be seen as a reachability problem where for an assumed initial state S^0 the state S^k, such that following transitions $S^0 \overset{i}{\rightarrow} S^k \overset{T_c-1}{\rightarrow} S^k$ holds, is sought. Of course, for any $S^k \in S_c$ the following property holds $S^k \overset{T_c-1}{\rightarrow} S^k$.

The introduced concept of state space allows to define the behavior of the SCCP system under the term of which is understood the set of all possible state spaces which are reachable in a given SCCP structure with different initial states and different DPR sets. To present the possibilities of the cyclic processes course control, let's consider the SCCP from Fig. 9.3b. Its state space following Table 9.2 is illustrated in Fig. 9.14a. In turn Fig. 9.14b shows another state space generated by the same initial processes allocation, however controlled by the slightly changed set of DPRs $\sigma_2 = (P_1, P_2)$, $\sigma_3 = (P_2, P_3, P_3)$, $\sigma_5 = (P_1, P_3, P_3)$. It is therefore easy to see that considered SCCP structure corresponds to the behavior which results (for the same initial state S_2) in at least two cyclic steady states of different cycles. In particular, this means that changing DPRs from $\sigma_2 = (P_1, P_2)$, $\sigma_3 = (P_2, P_3)$, $\sigma_5 = (P_1, P_3)$ (guaranteeing a cyclical course with a period of 9) in the S_2 state (corresponding to PA_2), to $\sigma_2 = (P_1, P_2)$, $\sigma_3 = (P_2, P_3, P_3)$, $\sigma_5 = (P_1, P_3, P_3)$

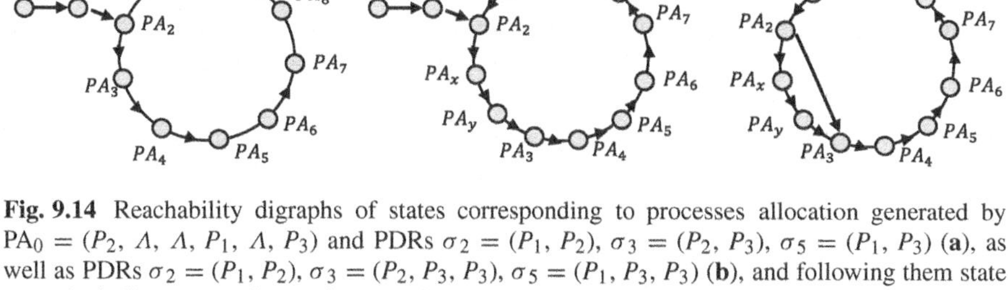

Fig. 9.14 Reachability digraphs of states corresponding to processes allocation generated by $\text{PA}_0 = (P_2, \Lambda, \Lambda, P_1, \Lambda, P_3)$ and PDRs $\sigma_2 = (P_1, P_2)$, $\sigma_3 = (P_2, P_3)$, $\sigma_5 = (P_1, P_3)$ (**a**), as well as PDRs $\sigma_2 = (P_1, P_2)$, $\sigma_3 = (P_2, P_3, P_3)$, $\sigma_5 = (P_1, P_3, P_3)$ (**b**), and following them state space including two cyclic steady states (**c**)

results in a cyclical course with a period $T = 11$, see Fig. 9.14c. Return to the previous run allows, in the S_2 state, a change restoring previously changed DPRs.

In that context, the multimodal processes that can be seen as processes composed of local cyclic processes lead to two fundamental questions: Does there exist a set of DPRs and an initial state enabling cyclic steady state of a given SCCP's structure while following requirements caused by multimodal processes at hand? Does there exist the SCCP's structure with a given initial state and a set of DPR such that an assumed multimodal processes flow can be achieved?

9.3.1.3 Periodicity Conditions

Periodicity or cyclicality shows a feature of the system whose behavior is described by repetitive changes (states) occurring at regular intervals. An example of this type of systems are cases of tram and bus lines running cyclically along closed loop routes stop at the depots they share. Similar networks form metro lines, suburban transport, and long-distance rail. An example of other objects in which the dominant role plays intralogistics solutions are manufacturing systems in which inter-operational transport is carried out through AGVS and/or milk-run systems. All these cases have the same feature which is the occurrence of cyclically recurring repetitive processes sharing commonly used system resources. The consequence of this is the need to guarantee their admissible, i.e., smooth (e.g., congestion and deadlock-free) flow.

9.3.1.4 Deadlock Prevention Conditions

Deadlock-freeness is a special property, stating that the system can never be in a situation in which no further progress is possible. In other words, such system's property supposes its indefinitely run. However, the problem of determining necessary and sufficient conditions the fulfillment of which guarantees the deadlock-free behavior of the system belongs to NP-hard problems.

In this study, assuming that the guarantee of deadlock-free system behavior provides a guarantee of its cyclicality we focus mainly on the problem of cyclic systems synthesis instead of searching for deadlock-avoidance control methods. Searching for appropriate polynomial computational complexity methods we focus on systems with regular topology structures. The acceptable solutions obtained in this way do not exhaust all possible ones. This means that the conditions under which they were obtained are just sufficient ones. They therefore correspond to the deadlock prevention conditions and not to deadlock avoidance conditions guaranteeing analysis of all possible cases (i.e., corresponding to necessary and sufficient conditions).

In order to present the proposed approach, let's consider supply network of grid-like structure obtained in the course of clusters aggregation following specific rules of fractal-like growth, see Fig. 9.6a. The grid-like structures are composed of

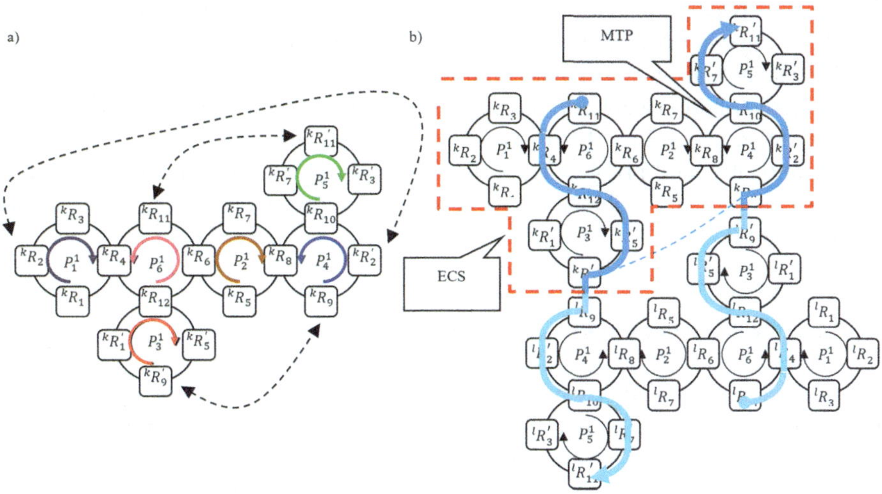

Fig. 9.15 Illustration of the covered form of ECS (**a**) and a fragment of the delivery network (**b**) made of ECS from Fig. 9.6c

clusters (involved in a fractal growth) which are identical repeating substructures, called Elementary Covering Structures (ECS), see Fig. 9.6c. More formally, an ECS is a connected digraph comprised of elementary substructures (modeling local cyclic processes) which occur in the regular structure of a distribution delivery network such that: the graph is composed of elementary cyclic digraphs which model local transportation processes, and can be used to tessellate a given regular structure.

Considered the "covered" form of ECS from Fig. 9.6c arises as a result of "gluing together" selected vertices of ECS, see Fig. 9.15a. Which particular vertices are "glued" together in the "covered" form is determined by the choice of those resources of the elementary transport structure which are shared with the resources of neighboring structures of the distribution network. For example, a vertex corresponding to resource ${}^{k}R_9$ is glued with a vertex corresponding to resource ${}^{k}R'_{11}$, because resource ${}^{k}R'_9$ is shared with resource ${}^{1}R_9$ which is a counterpart of ${}^{k}R_9$, see Fig. 9.15b.

It can be shown that if the traffic flow in a given covered form of an ECS is free of deadlocks, i.e., the schedule which specifies it is also a cyclic schedule, the flow of traffic in the entire transport network consisting of uncovered forms of ECSs also has a cyclic nature (Bocewicz & Banaszak, 2017; Bocewicz et al., 2017c, 2020, 2021).

The behavior of a given supply network shown in Fig. 9.15b can be predicted on the basis of the behavior of its ECSs. It is easy to observe that an initial allocation of local processes in a regular network (which maps the allocation of processes in the ECS) will be followed in each individual ECS by a next allocation of local processes in compliance with the same PDRs. That is because if one replicates the same initial process allocation in all remaining ECSs, the structure will be free of collisions between processes which use the same resources (both shared resources and those integrated/merged in the covered representation).

This means that cyclic behavior of the ECS implies cyclic behavior of the regular-structure distribution network, as well as multimodal transport processes (MTP) carried out in it, see Fig. 9.15b. Moreover it means that by solving a small-scale computationally hard problem (associated with an ECS), one can solve, in online mode, a large-scale problem associated with a corresponding regular-structure supply network.

9.3.1.5 Cyclic Scheduling

Many models and methods have been proposed to solve the cyclic scheduling problem in which some sequences of activities are repeated an indefinite number of times. Among them, the mathematical programming approach, max-plus algebra, evolutionary algorithms, and Petri nets frameworks belong to the more frequently used. Most of them are oriented at finding of a minimal cycle or maximal throughput while assuming deadlock-free processes flow (Crama et al., 2000; Dawande et al., 2005; Kats & Levner, 1998; Li et al., 2016; Wallace & Yorke-Smit, 2020).

Because declarative modeling (implemented in constraint logic programming environment) based approaches to cyclic scheduling are quite unique, let's show its implementation on example of the shop floor layout shown in Fig. 9.10. It consists of a warehouse R_w, a supermarket R_s, and five production cells R_1–R_5. The network of transport connections served by two tugger trains consists of a set of docking stations M_1–M_4 for tugger trains which deliver intermediate components from the warehouse to the production cells and a set of docking stations S_1–S_3 for tugger trains which pick up finished goods to supply them to the supermarket. By analogy to the milk-run schema, a multimodal process is understood here as a workstation-to-workstation production flow process, whose sections are local, cyclically repeated, milk-run tours.

As it is easy to notice, whether or not production takt time TP = 30 u.t. can be achieved is conditioned by timely (just-in-time) delivery/pickup of intermediate components/finished products to/from the given tugger train docking stations, see Figs. 9.10 and 9.17. In other words, the production flow schedule shown in Fig. 9.16 determines the schedule of visits to the individual tugger train docking stations.

In general case, the following question can be considered: Do there exist routes for the given tugger train fleet and batch size ρ, such that batches can be moved (delivered/picked up) along them to and from the given docking stations at time points determined by the production flow schedule from Fig. 9.16? By the size of production batch is further understood a number of jobs executed during one production cycle.

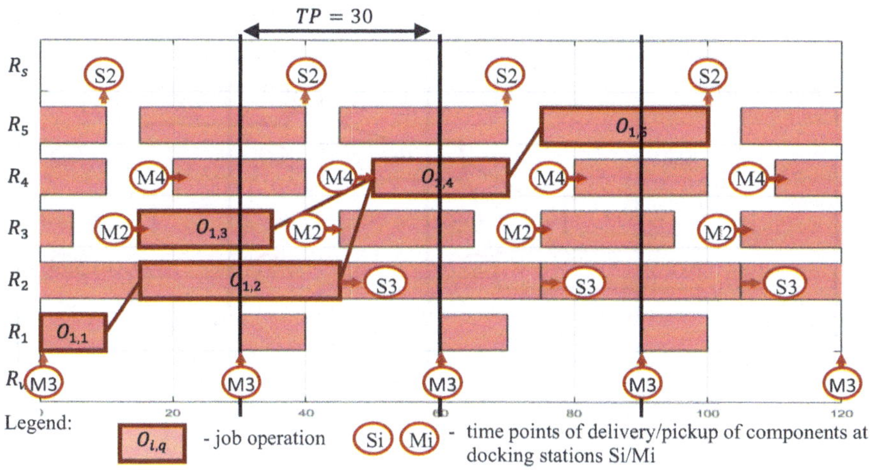

Fig. 9.16 A Gantt chart of production flow in a system from Fig. 9.10

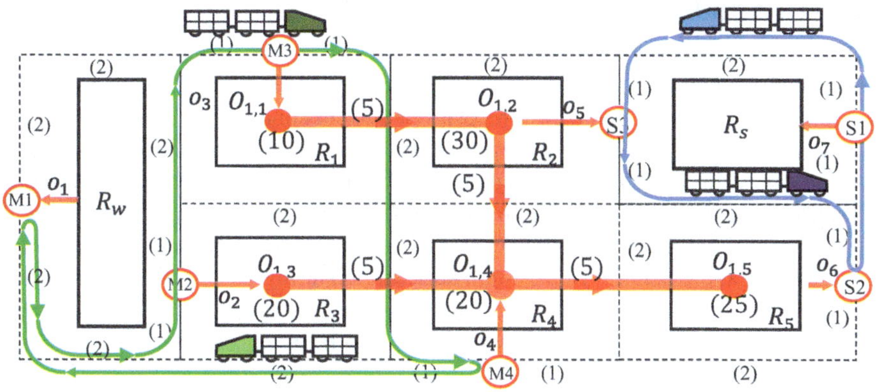

Fig. 9.17 Two separated transport routes each one followed by two tugger trains

Answer to the above question is provided by the one of feasible solutions shown in Fig. 9.17 where two cyclic routes one to be traveled by two tugger trains were established. A Gantt chart illustrating production flow in a system implementing this type of solution is shown in Fig. 9.18.

The solution involves the usage of a four tugger trains fleet assuming the size of the production batch equal to $\rho = 1$. There is however another solution (Bocewicz et al., 2019a) where the available fleet consists of two tugger trains following the same route while delivering production batches size of $\rho = 2$. It means that by increasing the size of the production batch, one can reduce the number of tugger trains used by a fleet, however, at the cost of reducing their utilization rate.

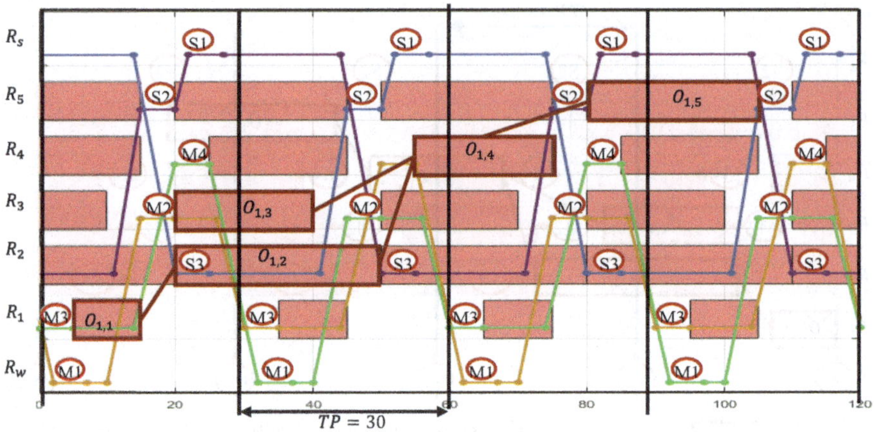

Fig. 9.18 A Gantt chart of production flow in a system from Fig. 9.17

9.4 Concluding Remarks

This chapter attempts an alternative approach to the synthesis of systems that guarantee the expected behavior. The class of systems under consideration includes cases which are often found in practice and characterized by the occurrence of concurrent cyclical processes implemented in them. Occurring in this area problems of analysis (searching for conditions that guarantee assumed system behavior with given constraints imposed by its structure) or synthesis (searching for conditions guaranteeing the existence of a system structure ensuring its assumed functioning) belong to the class of NP-hard problems. This means the need for a laborious and customized by the specificity of the instance, search for an approximate solution to the considered problem.

The presented approach can be seen as a kind of alternative to the above one. It assumes that the behavior of systems with regular structures is fully characterized by the behavior of its elementary structures. In other words—the behavior of a system established from standard, homogeneous recursively combined subsystems can be described using parameters specifying his constituent elements. This assumption is confirmed by the examples observed in practice examples of systems with grid and/or fractal structures (typical for the development of urban areas, modular solutions of electronic equipment structures based on the plug-and-go principle, and so on). In solutions of this type, the above-mentioned problems of analysis and synthesis boil down to the following questions: Does a given structure of the system guarantee its assumed behavior? Whether a given system behavior can be obtained in its structure composed in a regular way of given components? The computational complexity of problems formulated in this way allows one to solve them online in situations encountered in practice. This is possible due to the fact that the small scale of the component elementary problems despite the fact that they belong to the class of NP-hard ones can be resolved in online mode. This means, in turn, that the polynomial nature of computational complexity of a problem composed of recursively related local sub-problems allows for its quick solution.

Possibilities of using the presented paradigm boiling down to a phrase: "from knowledge of the nature of the elementary components of the system and their relationships the behavior of the entire (composed of them) system can be recreated" illustrated on selected examples of production systems and public transport. Declarative modeling paradigm used to describe the distribution networks with grid/fractal structures under consideration made it possible to formulate the synthesis problem of multimodal processes carried out in structures of concurrently flowing cyclic process. In that context, the proposed methodology aimed at assessment of possible mesh-like distribution networks carrying out multimodal processes flow is used in DSSs supporting the decision-maker in typical logistics management tasks. Examples illustrating such possibilities relate to the milk-run planning problems implied by routing, scheduling, and batching that arise during delivery of products to several production cells (Bocewicz et al., 2019a).

Of course, the discussed issues do not exhaust all related issues, as for example concerning the need to take into account the fuzziness of the data describing the modeled systems (Bocewicz et al., 2019b) as well as open problems. Relevant examples here are the problems of reachability (Bocewicz et al., 2014b, 2017a), and in particular the mutual reachability of cyclic steady states in the space of states achieved in a given system structure and resolvability of problems of cyclical processes synchronization which in integers domain can be classified as diophantine problems (Bocewicz et al., 2009).

References

Banaszak, A. Z., & Krogh, H. B. (1990). Deadlock avoidance in flexible manufacturing systems with concurrently competing process flows. *IEEE Transactions on Robotics and Automation, 6*(6), 724–734. https://doi.org/10.1109/70.63273

Banaszak, Z. (1997). Distributed bottleneck control for repetitive production systems. *Journal of Intelligent Manufacturing, 8*, 415–424. https://doi.org/10.1023/A:1018510300148

Banaszak, Z., & Polak, M. (2005). Concurrent processes flow prototyping. In A. Dolgui, J. Soldek, & O. Zaikin (Eds.), *Supply chain optimisation. Applied optimization* (Vol. 94). Springer. https://doi.org/10.1007/0-387-23581-7_7

Barakou, F., Koukoula, D., Hatziargyriou, N., & Dimeas, A. (2015). Fractal geometry for distribution grid topologies. In *2015 IEEE Eindhoven PowerTech*, Eindhoven, 2015 (pp. 1–6). https://doi.org/10.1109/PTC.2015.7232496.

Bocewicz, G., & Banaszak, Z. (2015). Multimodal processes scheduling in mesh-like network environment. *Archives of Control Sciences, 25*(2), 237–261. https://doi.org/10.1515/acsc-2015-0016

Bocewicz, G., & Banaszak, Z. (2017). Designing mass-customized network of passenger services subject to grid topology constraints. In R. Szewczyk, C. Zieliński, & M. Kaliczyńska (Eds.), *Automation 2017. ICA 2017. Advances in intelligent systems and computing* (Vol. 550). Springer. https://doi.org/10.1007/978-3-319-54042-9_11

Bocewicz, G., Banaszak, Z., & Nielsen, I. (2017b). Delivery-flow routing and scheduling subject to constraints imposed by vehicle flows in fractal-like networks. *Archives of Control Sciences, 27(LXIII)*(2), 133–148. https://doi.org/10.1515/acsc-2017-0008

Bocewicz, G., Banaszak, Z., & Nielsen, I. (2017c). Reduction of congestion in transport networks with a fractal structure. In Świątek et al. (Eds.), *Proceedings of 37th International Conference on Information Systems Architecture and Technology—ISAT 2016—Part III, Advances in Intelligent Systems and Computing* (Vol. 523, pp. 189–201). https://doi.org/10.1007/978-3-319-46589-0_15

Bocewicz, G., Banaszak, Z., & Nielsen, I. (2019b). Multimodal processes prototyping subject to grid-like network and fuzzy operation time constraints. *Annals of Operation Research, 273*, 561–585. https://doi.org/10.1007/s10479-017-2468-5

Bocewicz, G., Banaszak, Z., Nielsen, I., & Muszyński, W. (2017a). Re-scheduling of AGVs steady state flow. In *20th IFAC World Congress*. IFAC-PapersOnLine 50 (pp. 3493–3498). https://doi.org/10.1016/j.ifacol.2017.08.935.

Bocewicz, G., Banaszak, Z., & Pawlewski, P. (2014a). Multimodal cyclic processes scheduling in fractal structure networks environment. *IFAC Proceedings, 47*, 8939–8946. https://doi.org/10.3182/20140824-6-ZA-1003.01524

Bocewicz, G., Banaszak, Z., & Pawlewski, P. (2014b). Cyclic steady state space refinement. In R. Szewczyk, C. Zieliński, & M. Kaliczyńska (Eds.), *Recent advances in automation, robotics and measuring techniques. Advances in intelligent systems and computing* (Vol. 267). Springer. https://doi.org/10.1007/978-3-319-05353-0_2

Bocewicz, G., & Banaszak, Z. A. (2013). Declarative approach to cyclic steady state space refinement: Periodic process scheduling. *The International Journal of Advanced Manufacturing Technology, 67*(1–4), 137–155. https://doi.org/10.1007/s00170-013-4760-02013i

Bocewicz, G., Jardzioch, A., & Banaszak, Z. (2016). Modelling and performance evaluation of fractal topology streets network. In S. Omatu et al. (Eds.), *13th International Conference on Distributed Computing and Artificial Intelligence. Advances in Intelligent Systems and Computing* (Vol. 474). Springer. https://doi.org/10.1007/978-3-319-40162-1_52

Bocewicz, G., Muszyński, W., & Banaszak, Z. (2015). Models of multimodal networks and transport processes, 16. *Bulletin of the Polish Academy of Sciences Technical Sciences, 63*(3), 635–650. https://doi.org/10.1515/bpasts-2015-0075

Bocewicz, G., Nielsen, I., & Banaszak, Z. (2020). Blockage-free route planning for in-plant milk-run material delivery systems. In W. Bożejko & G. Bocewicz (Eds.), *Modelling and performance analysis of cyclic systems. Studies in systems, decision and control* (Vol. 241). Springer. https://doi.org/10.1007/978-3-030-27652-2_6

Bocewicz, G., Nielsen, I., & Banaszak, Z. (2021). Rerouting and rescheduling of in-plant milk run based delivery subject to supply Reconfigurability constraints. In G. Bocewicz, J. Pempera, & V. Toporkov (Eds.), *Performance evaluation models for distributed service networks. Studies in systems, decision and control* (Vol. 343). Springer. https://doi.org/10.1007/978-3-030-67063-4_4

Bocewicz, G., Nielsen, P., & Banaszak, Z. (2019a). Declarative modeling of a Milk-run vehicle routing problem for split and merge supply streams scheduling. In J. Świątek, L. Borzemski, & Z. Wilimowska (Eds.), *Proceedings of 39th International Conference on Information Systems Architecture and Technology – ISAT 2018. ISAT 2018. Advances in Intelligent Systems and Computing* (Vol. 853). Springer. https://doi.org/10.1007/978-3-319-99996-8_15

Bocewicz, G., Nielsen, P., Banaszak, Z., & Dang, V. Q. (2012). Cyclic steady state refinement: Multimodal processes perspective. In J. Frick & B. T. Laugen (Eds.), *Proceedings of the international conference on advances in production management systems* (Vol. 384: IFIP advances in information and communication technology, 1st ed., pp. 18–26). Springer. https://doi.org/10.1007/978-3-642-33980-6_3

Bocewicz, G., Wójcik, R., & Banaszak, Z. (2009). On undecidability of cyclic scheduling problems. In D. Karagiannis & Z. Jin (Eds.), *Knowledge science, engineering and management. KSEM 2009. Lecture notes in computer science* (Vol. 5914). Springer. https://doi.org/10.1007/978-3-642-10488-6_31

Bocewicz, G., Wójcik, R., & Banaszak, Z. (2018). Cyclic steady state behavior subject to grid-like network constraints. In S. Omatu, S. Rodríguez, G. Villarrubia, P. Faria, P. Sitek, & J. Prieto (Eds.), *14th International Conference on Distributed Computing and Artificial Intelligence. DCAI 2017. Advances in Intelligent Systems and Computing* (Vol. 620). Springer. https://doi.org/10.1007/978-3-319-62410-5_3

Bocewicz, G., Wójcik, R., Banaszak, Z. A., & Pawlewski, P. (2013). Multimodal processes rescheduling: Cyclic steady states space approach. *Mathematical Problems in Engineering, 2013*, 407096. https://doi.org/10.1155/2013/407096

Crama, Y., Kats, V., van de Klundert, J., et al. (2000). Cyclic scheduling in robotic flowshops. *Annals of Operations Research, 96*, 97–124. https://doi.org/10.1023/A:1018995317468

Dawande, M. W., Geismar, H. N., & Sethi, S. P. (2005). Dominance of cyclic solutions and challenges in the scheduling of robotic cells. *SIAM Review, 47*(4), 709–721. Retrieved September 3, 2021, from www.jstor.org/stable/20453702

Kamoun, H., & Sriskandarajah, C. (1993). The complexity of scheduling jobs in repetitive manufacturing systems. *European Journal of Operational Research, 70*(3), 350–364. https://doi.org/10.1016/0377-2217(93)90247-K

Kats, V., & Levner, E. (1998). Cyclic scheduling of operations for a part type in an FMS handled by a single robot: A parametric critical-path approach. *International Journal of Flexible Manufacturing Systems, 10*, 129–138. https://doi.org/10.1023/A:1008045414118

Levner, E., Kats, V., López, A., de Pablo, D., & Cheng, T. C. E. (2010). Complexity of cyclic scheduling problems: A state-of-the-art survey. *Computers & Industrial Engineering, 59*(2), 352–361. https://doi.org/10.1016/j.cie.2010.03.013

Li, Q., Pogromskiy, A. Y., Adriaansen, A. C., & Udding, J. T. (2016). A control of collision and deadlock avoidance for automated guided vehicles with a fault-tolerance capability. *International Journal of Advanced Robotic Systems, 13*(2), 64. https://doi.org/10.5772/62685

Majdzik, P., & Obuchowicz, A. (2006). Prototyping of concurrent cyclic processes. In *IFAC Proceedings. 3rd IFAC Workshop on Discrete-Event System Design* (Vol. 39, pp. 257–262). https://doi.org/10.3182/20060926-3-PL-4904.00043

Montreuil, B. (1999). Fractal layout organization for job shop environments. *International Journal of Production Research, 37*(3), 501–521. https://doi.org/10.1080/002075499191643

Polak, M., Majdzik, P., Banaszak, Z. A., & Wójcik, R. (2004). The performance evaluation tool for automated prototyping of concurrent cyclic processes. *Fundamenta Informaticae, 60*(1–4), 269–289.

Polak, M., Wojcik, R., Majdzik, P., & Banaszak, Z. (2003). Prototyping of distributed control procedures in concurrent cyclic processes systems. In *Proceedings EFTA 2003. 2003 IEEE Conference on Emerging Technologies and Factory Automation (Cat. No.03TH8696)* (Vol. 2, pp. 261–268). https://doi.org/10.1109/ETFA.2003.1248709

Venkatadri, U., Rardin, R. L., & Montreuil, B. (1997). A design methodology for fractal layout organization. *IIE Transactions, 29*, 911–924. https://doi.org/10.1023/A:1018540000690

Wallace, M., & Yorke-Smit, N. (2020). A new constraint programming model and solving for the cyclic hoist scheduling problem. *Constraints, 25*, 319–337. https://doi.org/10.1007/s10601-020-09316-z

Wójcik, R., Bocewicz, G., & Banaszak, Z. (2016). Towards cyclic scheduling of grid-like structure networks. In L. Borzemski, A. Grzech, J. Świątek, & Z. Wilimowska (Eds.), *Proceedings of 36th International Conference on Information Systems Architecture and Technology – ISAT 2015 – Part I. Advances in Intelligent Systems and Computing* (Vol. 429). Springer. https://doi.org/10.1007/978-3-319-28555-9_2

Chapter 10
A General Method for Designing General Systems

Seiji Kurosu

Abstract When using Work Design as a system design method, there are two main aspects to how a system is perceived. One is the aspect of thinking of the system as an abstract and insubstantial product of thinking. The other is the aspect as a concrete entity consisting of things and human beings. In this chapter, the system as an entity and the system of thinking are described and organized as formally as possible. It clarifies the system concept used in Work Design. Work Design is one of the general system design methods for general systems, developed by Dr. Nadler. Specifically, the system concepts used in Work Design are classified into four types: purpose expression system, input-output expression system, procedure expression system, and catalyst expression system. The purpose expression system, the input-output expression system, and the procedure expression system are considered to be thought systems. The catalyst expression system is considered to be a real system. It is stated that system design is to realize a system from a thought system to a real system. In addition, the features of the system design method by Work Design will be organized.

Keywords System · Design · Method · Formalization

10.1 Introduction

Work Design is considered to be a general system design method for general systems, developed by Nadler (1963). When using Work Design as a system design method, there are two major aspects to the way systems are perceived. The first is to think of a system as an abstract and insubstantial product of thought. The other is as a concrete entity consisting of objects and people. In the process of system design, it is necessary to draw the system in thought first, and then build up the system as

S. Kurosu (✉)
Waseda University, Tokyo, Japan
e-mail: cross@waseda.jp

K. Kijima et al. (eds.), *Systems Research I*, Translational Systems Sciences 26,
https://doi.org/10.1007/978-981-16-9937-5_10

an entity based on it. Then, it is necessary to make a smooth transition from the first phase to the second phase. If we confuse the two phases, we may overlook the potential of the system design itself.

For example, assume a car headlamp. Then consider designing a "system that brightens the front." The form of the system will be an input-output system. In an input-output system, input and output are usually set first, and then the system is considered. In this case, what should be set as the input of the "system that brightens the front"? What should be set for the output?

Suppose you think of the input as "a front surrounded by darkness" and the output as a "bright front." Then, the "front surrounded by darkness" enters the system, and the "bright front" comes out of the system. However, is it possible to design a system that contains something called "the front surrounded by darkness"? Also, is it possible to design a system that outputs something called "brightened front"? What kind of system is the system that outputs "bright front" when "darkness in front" is input?

The above example seems to say something like this: "The inputs of the system are usually things and people, and we cannot think of anything like darkness." Certainly there seems to be some sense of incongruity. So should darkness be treated as an input? If so, how should darkness be treated? The background of this idea may be that the system is regarded as a box (machines, equipment, factories, schools, offices, etc.). Perhaps it is because they are trying to understand the system as an entity from the beginning. In this chapter, from the viewpoint of general system theory, the system as an entity and the system in thinking are formally described and organized.

Then, based on the formal system, the system concept used in Work Design is clarified. Specifically, the system concept is classified into four categories in the order of design process: purpose expression system, input-output expression system, procedure expression system, and catalyst expression system. In particular, the two system concepts of the purpose expression system and the catalyst expression system are clarified. In the system design process by Work Design, the first three are considered as thinking systems, and the remaining one is considered as a real system. Then, we conclude that system design is to build a system from a thinking system to a real system. In addition, by clarifying the system designed by Work Design, the system design by Work Design can be applied more effectively.

In the following, the term "system design" means system design exclusively by Work Design.

10.2 Formalization of System Concept

To begin, we will attempt to situate the concept of basic systems in Work Design within the framework of general systems, which is the most basic of all systems concepts. To do so, we will first introduce a formal general systems theory.

10.2.1 General Systems

Many definitions of systems have been attempted. Perhaps the most basic and general of these definitions is that "a system is a set of elements and their relation" (Mesarovic & Takahara, 1975; Klir, 1991). Mesarovic and Takahara (1989) further defines it formally as follows. A relation is a subset of a direct product set. However, a direct product set is a set of ordered pairs (x, y) created by taking any elements x, y from each of the two sets A and B. This is denoted as $A \times B$.

For example, if $A = \{a_1, a_2\}$ and $B = \{b_1, b_2\}$, then the direct product set

$$D = A \times B \text{ is } D = \{(a_1, b_1), (a_1, b_2), (a_2, b_1), (a_2, b_2)\}$$

Then, a relation is defined as a subset of a set D that collects the appropriate elements under certain rules. In other words, let R be a relation

$$R \subset D$$

More generally, let V_j ($j = 1, 2, \ldots, n$) as a set, it is as follows.

$$R \subset V_1 \times V_2 \times \cdots \times V_n \tag{10.1}$$

This relation R is the general form of the system.

10.2.2 Input-Output System

Equation (10.1) can be rewritten as an input-output system if we consider that the first k of V_j act as and the remaining n-k of V_j act as the result (Mesarovic & Takahara, 1989). Therefore, we define it as follows.

Definition 10.1 A system S denoted by

$$S \subset I \times O \tag{10.2}$$

is called an input-output system. Let $I = V_1 \times V_2 \times \cdots \times V_k$ and $O = V_k + 1 \times V_k + 2 \times \cdots \times V_n$.

For example, suppose we have a set I = {blue paper, yellow paper}. On the other hand, suppose there is a set J = {type☆, type∘}. Then, the direct product set L of I and J is {blue☆, blue∘, yellow☆, yellow∘}. However, the words "paper" and "type" have been omitted here. Next, from the direct product set of L, discard blue∘ and yellow☆ to create a subset of L, {blue☆, yellow∘}. Let {blue☆, yellow∘} be O. Then, if we make the direct product set of I and O, we get $I \times O$ = {blue, yellow} × {blue☆, yellow∘}. If we denote the subset {(blue, blue☆), (yellow,

yellow∘)} of the direct product $I \times O$ as S_G, we can think of S_G as an input-output system of $I \times O$.

$S_G = \{(\text{blue}, \text{blue}☆), (\text{yellow}, \text{yellow}∘)\} \subset \{(\text{blue}, \text{yellow})\} \times \{(\text{blue}☆, \text{yellow}∘)\} = I \times O$

In other words, S_G shows the relation between the set I, which consists of blue paper and yellow paper, and the set O, which consists of blue☆ and yellow∘. And if we interpret the set O consisting of blue☆ and yellow∘ as Christmas decorations, S_G shows the relation between the input of blue paper and yellow paper and the output of Christmas decorations. However, S_G does not show that the Christmas decorations were made from the blue and yellow papers. In Sect. 10.2.3, we will consider how to create such a representation by following four steps.

10.2.3 Purpose Expression System

Equation (10.2) shows the relationship between input and output, but it does not clearly indicate the direction from input to output. Therefore, we can express the correspondence from input I to output O using a mapping as follows:

$$F : I \to O \tag{10.3}$$

In this way, we can clearly express the direction from I to O. Here, this direction is from cause to effect. Then, F in Eq. (10.3) can also be mapped to the purpose expression in Work Design. This is because, when we apply the purpose expression in Work Design defined by the author, it may be re-expressed as "converting (input) into (output)." For example, the purpose expression "Make Christmas decorations {blue ☆, yellow ∘}" can be rewritten as follows.

"convert blue and yellow paper into Christmas decorations {blue☆, yellow∘}.
Input: blue and yellow paper
Output: Christmas decorations {blue☆, yellow∘}

This means that the correspondence is isomorphic to the meaning of Eq. (10.3), "F is the map with which the first set I and the last set O are associated." In other words, it is as shown in Fig. 10.1.

	Mapping
Mathematics	I → O
↓	↓
Work design	I → O
	Purpose

Fig. 10.1 Relationship between mapping and purpose

A mapping can be said to be a function. Therefore, a mapping is also a system. Therefore, we can think of a purpose expression as one representation of a system. From this perspective, we define it as follows.

Definition 10.2 A sentence expression that can be rewritten as "transform ˜(*I*) into … (*O*)" is a purpose expression system S_F.

According to Definition 10.2, a sentence expression that cannot be rewritten as "convert ~(*I*) into … (*O*)" cannot be called a purpose expression system.

For example, "walking" and "eating" would not be called purposes in Work Design, because they represent states, not outcomes achieved by "converting ~(*I*) into … (*O*)". However, if the expression is "make the patient walk," for example, this can be rewritten as "convert the patient (*I*) into a walking patient (*O*)," so this sentence is a purpose expression system. In addition, the expression "make the patient walk" can be interpreted as a purpose because it accomplishes a purpose and produces results in Work Design. Therefore, the mapping F in Eq. (10.3) can be interpreted as corresponding to the purpose expression system in Work Design in many cases.

However, in the process of designing a system by Work Design, especially in the process of expanding purposes, there are cases where the inputs are not clearly expressed. For example, in the case of "manufacture automobiles," the part corresponding to the input is not expressed.

This is because at this stage, the focus is on the thinking system, and the input does not have to be the input as a real system. The inputs of the real system can be considered in later stages by creative idea generation to come up with a number of alternatives. Therefore, it is interpreted as a deliberate lack of clarity.

In Work Design, however, inputs are sought and clarified in the following design process. Therefore, although "manufacture a car" is incomplete, it can be treated as a purpose expression system in the system design process.

10.2.4 Procedure Expression System

Although Eq. (10.3) represents the macroscopic correspondence oriented from input to output, it does not represent each detailed and specific way of creating, or converting, input to output. There are various ways to create a star-shaped blue paper and a circle-shaped yellow paper from blue and yellow paper. Cutting out with scissors is probably the most common. Another method is to make star-shaped and circle-shaped die cutters and use them to punch out the paper. Other methods include burning or melting the unnecessary parts other than the star and circle shapes.

The methods of cutting out, punching out, burning, melting, etc. are called the conversion procedure T in this chapter. In Work Design, this is called Sequence (Nadler, 1981). This T is the embodiment of F. Although the details have not yet been specified, it is an entity system with concreteness. However, there is more than one type of T, as shown in the example above.

On the other hand, F is unique. This is the point of difference between T and F. We can say that T is also a system. We will now define this system.

Definition 10.3 A system in which the transformation from input I to output O is expressed in terms of procedures is called a procedure expression system S_T.

10.2.5 Catalyst Expression System

There is a method of using scissors as a method of "cutting out" which is a conversion procedure, and there is also a method of using a knife. Furthermore, both scissors and knives have their own unique uses. Using a knife with scissors does not work. A knife has a way of using a knife. Further, these operations may be performed by humans or by machines.

From these facts, it is considered that the conversion method consists of procedures such as cutting, punching, burning, and melting, tools such as scissors and knives, knowledge of how to use them, and humans and machines that work. The tools and machines to be used are called the physical catalyst C_p, the method of using them is called the information catalyst C_k, and the human doing the work is called the human catalyst C_m.

Then, the conversion method can be thought of as consisting of four parts: the procedure, the physical and information catalysts, and the human catalyst.

It can be said that the catalyst is an element that is decided after the procedure is decided. Alternatively, it can be said that the catalyst is determined by refining the procedure. In other words, the catalyst is dependent on the procedure. Alternatively, a procedure can be regarded as a set of catalysts. In the above example, when the procedure of cutting out is decided, the catalyst of scissors or knife is decided. At the same time, we need an information catalyst, which is how to use the scissors or knife.

Information catalysts are used in connection with the procedure of cutting out. However, even with these, the conversion from input to output will still not really work. We will need a mutual relationship diagram showing when and how to use scissors and other catalysts. This diagram shows the relationships among the catalysts to convert the input of paper into the output of a star shape by cutting it with scissors.

We will denote the linkage diagram by Γ_C. C and Γ_C are used in combination to make sense. So we will put them in parentheses and represent them by S_H. That is

$$S_H = (C, \Gamma_C)$$

This S_H is the content of the system that converts the input as an entity into the output as an entity. So, let us denote this transformation as i (set) for the input as an entity and o (set) for the output as an entity

$$o = H(i)$$

where the input as entity is i (set) and the output as entity is o (set).

Here, I in general system theory that has been used so far is an input, I is an input in thinking. Let us now introduce a mapping δ_i, which serves to concretize i in an entity from I in thinking. That is

$$\delta_i : I \to i$$

In the same way, let us introduce a mapping δ_o, which is responsible for concretizing o as an entity from O in thought.

$$\delta_o : O \to o$$

The relationship between I, O, F, H, i, o, S_c, δ_i, δ_o are shown in Fig. 10.2.

The above rewrite is mathematically equivalent to the refinement of S_F into a more concrete function form called S_H. In this sense, S_T is also a concrete function form of S_F. In other words, based on S_F, we have found a rather rough function form called S_T. Then, according to T, we have found a concrete and substantive function form called S_H. Let us now define the catalyst expression system.

Definition 10.4 Conversion from input i as an entity to output o as an entity

$$o = H(i)$$

is represented by an appropriate set of catalysts C and their linkage schemes Γ_C

$$S_H = (C, \Gamma_C)$$

is called a catalyst expression system.

A catalyst expression system is a real system, but it does not necessarily have to have things or people in it. Computer software is interpreted here as an entity system

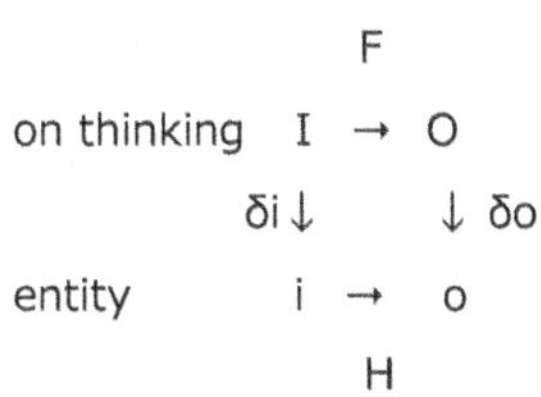

Fig. 10.2 Relationship between the entity system and the system on thinking

that consists only of information. This is also included in the catalyst expression system.

However, in Work Design, the information catalyst is also an entity, and it is handled differently depending on what specific form the information takes, such as document, voice, electrical signal, disk, etc.

10.3 System Design Process

Using the above system concept, let us organize the process of system design by Work Design.

The first step in system design by Work Design is to clarify the purpose. The purpose is expressed in sentences. An example of Christmas decorations is "Make ☆shaped blue paper and ○shaped yellow paper for Christmas decorations."

At this first stage, the outline of the input and output of the system to be designed was clarified. This is because the purpose expression shows, albeit implicitly, inputs and outputs. In the case of this purpose expression, the outputs are ☆shaped blue paper and ○shaped yellow paper for Christmas decorations. Then we know that the inputs are blue paper and yellow paper, or just white paper and blue and yellow paints. The output is immediately obvious from the purpose expression alone, but the input is not yet clearly defined at this stage. Therefore, in the second stage of Work Design, the inputs are clearly determined. The output will also be confirmed at the same time for confirmation. In other words, the second stage of Work Design is the determination of inputs and outputs. After this stage, the inputs and outputs are fixed for the time being. In actual Work Design, though, the inputs and outputs may change retroactively if another creative idea comes up in the third stage or later.

Stage 1: Purpose expression system

"To make ☆shaped blue paper and ○shaped yellow paper for Christmas decorations."

Stage 2: Input-output expression system

In Work Design, the system is represented by a hopper, not a box (Fig. 10.3).

Stage 3: Procedure expression system S_T

The third step in the Work Design is to find a procedure for converting input to output. From this stage, the box, or hopper, is filled in. For example, if we use the procedure of "cutting out" to convert input into output, the system at this stage can be called a "cutout system" (Fig. 10.4).

The hopper in the third stage is still in thought, but not yet in entity.

Stage 4: Catalyst expression system

The procedure of the fourth stage is to find out the catalysts necessary to realize the procedure and clarify their relational diagram so that the conversion from input to output can be concretely performed individually.

As mentioned earlier, for example, let us say you have decided to use a catalyst called scissors. Then, the procedure at this stage is to create a detailed relational

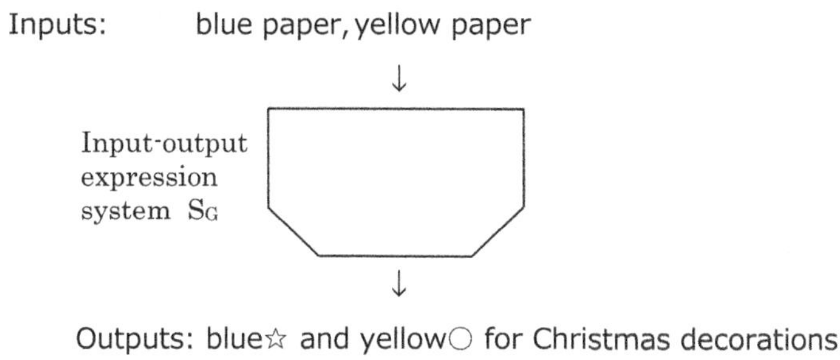

Fig. 10.3 Input-output expression system

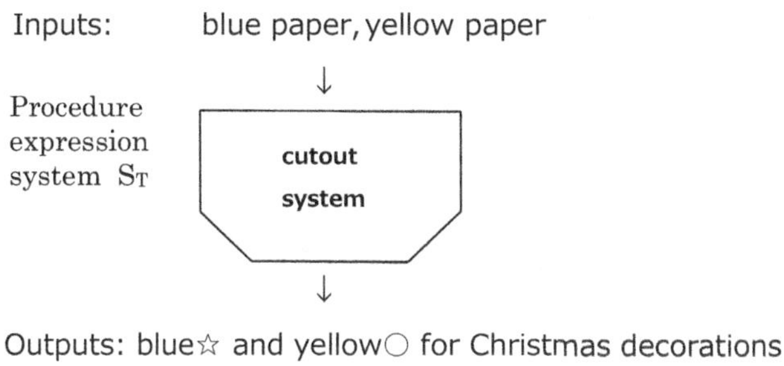

Fig. 10.4 Procedure expression system

diagram from the paper to the paper cut into a ☆ shape by the scissors and the information catalyst about how to use them. This example is so simple that anyone can do it without having to think about the details. However, if we were to have a robot do this work as an ideal system, for example, we would have to create a linkage diagram in which the robot would hold a piece of paper in one hand, hold a pair of scissors in the other, and cut the paper while moving the handle of the scissors up and down. For the first time, this system can convert the input as an entity into the output as an entity in every detail (Fig. 10.5).

Note that the catalyst expression system is an entity system.

By the way, we often write the system in a box or hoper as shown above. However, as we have seen here, the purpose expression system is insubstantial. Also, in the input-output expression system, we can use a box, but it will be blank inside. For the first time in the third stage, we worked on embedding procedures and other information inside the box. For the first time in the third stage, we worked on embedding procedures and other information in the hopper. However, that does not mean that the system has emerged in concrete detail in the third stage. From the stage of the purpose expression system, we recognize the system, though it is not clear. As it goes through the stages, it becomes clear and concrete. Then, in the fourth stage, the system is expressed by real entities (including physical phenomena), people, knowledge, etc. and their related diagrams.

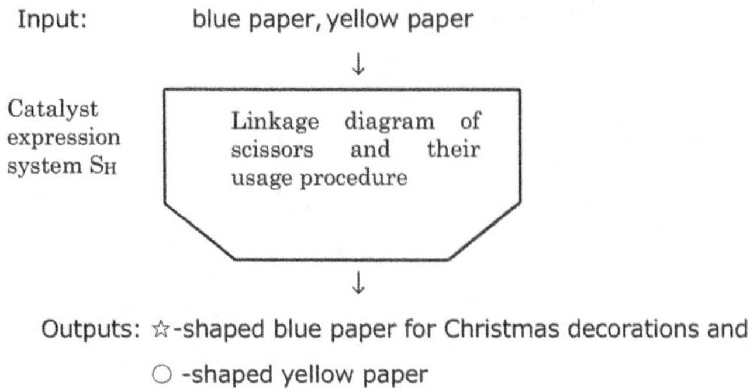

Fig. 10.5 Catalyst expression system

To briefly describe the above, it can be said that what was initially an expression based on inputs and outputs has been transferred to an expression based on a conversion method. In other words, the system design by Work Design can be roughly divided into the work of clarifying the input and the output and the work of devising the conversion method from the input to the output.

10.4 Significance of System Concept Clarification

10.4.1 System to Be Designed

The work of clarifying inputs and outputs is called "*I*/*O* clarification" in this chapter for convenience, and the work of proposing how to convert inputs to outputs is called "*I*/*O* conversion proposal."

In general, when we talk about system design, there are two ways of referring to it: one is to refer only to the work of *I*/*O* conversion planning, with *I*/*O* as a given, and the other is to refer to the work including *I*/*O* clarification. The former is called "system design in the narrow sense," and the latter is called "system design in the broad sense".

The author has always thought that the scope of the target system is unclear when referring to system design by Work Design. In other words, does the system design refer only to the inside of the hopper in the figure, or does it include input and output as well as the inside of the hopper? The object of system design in a narrow sense should be inside the hopper. However, in system design by Work Design, the clarification of inputs and outputs is also included in the system design.

10.4.2 Significance of Purpose Expression System

In this chapter, we have defined the purpose expression system. Using this definition, we can more clearly determine whether an arbitrary textual expression is a purpose expression or not, which was previously ambiguous. In other words, any expression that cannot be rewritten as "converting (input) into (output)" is not considered to be a purpose expression in general systems theory. And a textual representation that is not a purpose expression cannot be the target of system design. For these reasons, system design by Work Design starts with a representation of purpose. We cannot start with a general definition of a system, as described in Sect. 10.2.1, "A system is a set of elements and the relationships among them".

For example, a "car" is a system according to the general definition of a system. This is because an automobile is made up of parts and the relationships between them. However, in Work Design, this is not the only way to design an automobile system. The unique specifications of the automobile to be designed need to be identified. It is also necessary to identify the purpose of the automobile. It is possible to design a car (or something like a car) with a purpose expression system that says "to transport people to where they want to go".

But a system designed by Work Design, of course, fits into the general definition of a system.

10.4.3 Significance of Catalyst Expression System

Traditionally, the definition of a system in Work Design has been perceived as ambiguous by novices. As a result, some have argued that a catalyst is not a component of a system. It was also unclear which part of the system was considered to be the system. However, this chapter has clarified that the system in Work Design is the catalyst. It also provides a theoretical basis for clarifying the distinction between input and catalyst, which has been a problem in the past.

10.4.4 Interpretation of the Opening Problem

Let us now revisit the problem at the beginning of this chapter. "Darkness" certainly does not seem to be a thing or a person. However, at least in the input-output and procedure expression systems, there is no problem in entering "darkness" as it is. Also, the purpose expression system and the input-output expression system are just considering the correspondence between input and output. The problem is when that input is used as input to a system, such as a catalyst expression system.

A system may be thought of as something that transforms and processes the input. From this perspective, darkness is not an input because it is not subject to transformation or processing. Therefore, one might think that a system that satisfies the objective of "brightening the front" cannot be designed by Work Design. However, the author does not think so.

Instead of using the expression "transform" or "process," let us use the word "change," "rephrase." By doing so, you can turn darkness into lightness, which is a good input. Furthermore, a system for "brightening the front" can be designed using the concept of Work Design. For example, the system idea of "illuminating the front with a flashlight" would be one feasible idea. Moreover, the flashlight is a system of catalyst expression as well as a system of reality.

If the system can be designed with Work Design in this way, it is better to consider that there is a problem with the idea that "a system is something that converts and processes input". The expression "transform" or "convert" the darkness is unnatural, but "change" the front to a brighter front is not so unnatural.

In the above way, the problem of darkness can be solved without any discomfort. There also seems to be no need to distinguish it from situational input and so on. Furthermore, it should be clear that we should not think of systems as boxes (machines, devices, factories, schools, offices, etc.).

10.5 Conclusion

Summarizing the above, the following conclusions can be drawn.

1. The system concept used in Work Design was positioned within the framework of general systems, which is considered to be the most basic of all system concepts. This theoretically clarified the concept and procedure of system design by Work Design, which is one of the general system design methods.
2. From the viewpoint of general systems theory, there are two system concepts explicitly treated in Work Design: input-output expression system and procedure expression system, but this chapter adds two more concepts: purpose expression system and catalyst expression system.
3. From (1), it is clear that a purpose expression can basically be a system expression in which outputs and inputs are included in a sentence. However, in some purpose expressions, inputs and outputs are not clearly expressed, but clarifying them during design is one of the processes of system design by Work Design. By doing so, it is possible to clearly determine which purpose expressions are eligible for system design and which are not. This makes it a little easier to do purpose expansion that were previously said to be difficult.
4. By clarifying the concept of the catalyst expression system, we were able to clarify the characteristics of the realization of the system at each stage, which were previously ambiguous. As a result, the distinction between input and catalyst became theoretically clear.

5. System design by Work Design often consists of two major processes. The first is "clarification of *I*/O" and the second is "proposal of *I*/*O* transformation." The objects of Work Design are the inputs and outputs in "*I*/*O* clarification" and the inside of the box in "*I*/*O* transformation proposal." However, inputs and outputs do not fall within the scope of the system. The system is only a "box."

Recognizing the above, we can proceed with system design by Work Design while keeping the key points in mind.

References

Klir, G. J. (1991). *Facets of systems science*. Prenum Press.
Mesarovic, M. D., & Takahara, Y. (1975). *General systems theory: Mathematical foundations*. Academic Press.
Mesarovic, M. D., & Takahara, Y. (1989). *Abstract systems theory*. Springer.
Nadler, G. (1963). *Work Design*. Richard D. Irwin.
Nadler, G. (1981). *The planning and design approach*. Wiley.

Chapter 11
Modeling Complex Systems and Their Validation—General System Theoretical Approach

Naoki Shiba

Abstract This chapter discusses the methodological foundation of agent-based model (ABM) validation by presenting formal definitions of several related concepts. Simulation with agent-based models is often used as a method of complex systems research. We formalize several concepts related to model validity for agent-based social simulations from two standpoints, i.e., empiricism and constructivism, using homomorphism between systems according to the mathematical general systems theory (MGST). Based on the formal definitions of several validity concepts, we argue that the validity discussion in the ABM can be captured as the mixture of two validity concepts, i.e., the empirical validity and the constructive validity.

Keywords Model validation · Agent-based models · Complex systems

11.1 Introduction

Agent-based modeling is an approach to complex systems, especially in social simulation. Agent-based modeling is a promising method for the interdisciplinary exploration bridging the gap between social science and engineering. In this chapter, we discuss the methodological foundation of agent-based model validation. This issue has been discussed in several related research fields from several standpoints (Naylor and Finger, 1967; Law, 2007; Kueppers and Lenhard, 2005; Sargent, 2011; Oreskes et al., 1994; Grimm et al., 2005; Fagiolo et al., 2007; Sun and Tesfatsion, 2007; Moss, 2008; Boero and Squazzoni, 2005; Werker and Brenner, 2014; Morone and Taylor, 2010; Railsback and Grimm, 2019).

Naylor and Finger discussed three historical methods of validation (Naylor and Finger, 1967), and Law discussed model validation (Law, 2007). Kueppers and Lenhard presented a new perspective on the validity of a computer simulation

N. Shiba (✉)
College of Industrial Technology, Nihon University, Narashino, Chiba, Japan
e-mail: shiba.naoki@nihon-u.ac.jp

K. Kijima et al. (eds.), *Systems Research I*, Translational Systems Sciences 26,
https://doi.org/10.1007/978-981-16-9937-5_11

model (Kueppers and Lenhard, 2005). Sargent's tutorial gives a list of studies that are relevant to the modeling of social systems (Sargent, 2011). Oreskes et al. presented a relative perspective about model validation from the point of view of Earth scientists (Oreskes et al., 1994). Grimm et al. emphasized the qualitative aspect of model validation process (Grimm et al., 2005). Some researchers in computational economics sought to lay the foundation and to apply empirical validation (Fagiolo et al., 2007; Sun and Tesfatsion, 2007). Moss presented a critique to their empirical view of validation from a different perspective (Moss, 2008). Various researchers presented a taxonomy of agent-based models and differentiated the validation depending on the taxonomy (Boero and Squazzoni, 2005). Werker et al. discussed the method of model calibration from the standpoint of critical realism (Werker and Brenner, 2014). Morone et al. presented methods of model validation called docking and connecting, which are performed by comparing models (Morone and Taylor, 2010). Rainsback proposed a standard protocol to document models that enable the reader to understand models easily (Railsback and Grimm, 2019). However, the validity of the models has been a controversial topic in the agent-based social simulation field. One reason for this is that there is no consensus on the concepts related to model validation, especially in the community of agent-based social simulation researchers and practitioners (Arnold, 2019).

This chapter introduces formal definitions of several concepts related to model validation in agent-based social simulation. Without a rigid definition of key concepts, we can neither have a fruitful discussion nor reach a consensus between different perspectives to questions such as "What is validation?", "Is traditional and conventional empirical validation sufficient for agent-based models?", and "If traditional and conventional empirical validation is not sufficient, then how do we validate agent-based models?". The objective of this chapter is to provide a framework for this type of discussion.

For this purpose, we focus on the methodological discussion of Galliers in the information systems research field (Galliers, 1991). General system theory has been applied to information systems research (Takahara and Liu, 2006), and social simulation is an effective method for information systems research (Shiba, 2019). The discussion of Galliers helps us to grasp the organized perspective of a number of research methods, including social simulation for information system research. First, we interpret his methodological discussion, which is intended to cover information systems research (ISR) in general, in the context of social simulation with agent-based modeling. Then, with the support of the discussion of Galliers, we formalize the model validation activity from two standpoints, i.e., empirical and constructive views, within mathematical general systems framework. Finally, the following statements are assumed:

1. The validation method for agent-based models can be placed on the spectrum with two end points.
2. Researchers using agent-based modeling should disclose their model in as much detail as possible.

11.2 Information Systems Research (ISR) and Simulation

As described in the previous section, agent-based social simulation models have a wide variety in terms of approaches and methods. However, there appears to be no consensus concerning this variety. As a starting point to formalizing the validity of these approaches and methods, we introduce a methodological discussion in social science, especially in information systems research. Social science spans a wide range of studies, from purely human-oriented approaches to hybrid approaches for various human and technological aspects. Information systems research is typical of the latter approach and covers a broad range of topics and, as a consequence, methods of exploration are highly diversified. Furthermore, as one can infer from the fact that information systems research includes information and communication technology as its research target, there are many researchers in the information systems research community who have adopted computerized social simulation as their research method. Therefore, methodological discussions that have taken place in the information systems research community explicitly consider simulation as a possible research method, and the discussions can serve as a good guide for our effort to formalize the validity concept in agent-based models.

Information systems (IS) is a comprehensive research field in social science, covering the technological aspect of systems that process information and discusses broad areas of our society, its organization, and its management with a core focus on information (Mingers and Stowell, 1997; Lee, 1991).

Hence, the IS field comprises several traditional academic disciplines, has intrinsic interdisciplinary characteristics, and is lacking a dominant theory or a unified and specific method to cover the entire field systematically. As a result, several methods are used in ISR with several philosophical backgrounds.

In the methodological discussion of ISR, previous studies have considered the empirical approach and the constructive approach (Galliers, 1991; Vogel and Wetherbe, 1984; Farhoomand, 1987; Hamilton, 1982).

In the case of the empirical approach, we consider our experiences, which are gathered through our perceptions, as an epistemic basis. Based on these considerations, we have the following assumptions. Objects may exist independent of our recognition, i.e., the world exists regardless of our recognition (realism). In addition, as subjects, we can address questions about objects in a neutral and objective manner and can obtain answers to questions without distorting objects (epistemology based on subject–object dualism). Empiricists claim that this is exactly the strength of objective scientific methods. Many of the traditional methods used in natural science can be categorized into this approach. In particular, the positivist method, in which hypotheses are deductively derived from theories and should be validated experimentally or through observations, is typical of such an approach. There is a relaxed view of dualism accepting the theory—ladenness of observation, i.e., answers to questions about objects can be influenced by the theories of observers. However, empiricist methods share the common view that the "reality" of objects should be sought through our observations.

In the case of the constructive approach, we do not think that the meaning of social existence exists regardless of observers, as empiricists or positivists assert, but rather think that the meaning is generated through intersubjective dialogue between an originator of text and an interpreter of text (Hirschheim et al., 1995). The term "text" is used with the original meaning of the "woven" material and is an expression of social existence in some form. Research papers are typical texts. Constructivists claim that dialogue takes place via a paper as text, and the meaning is generated through the dialogue between originators (authors) and interpreters (readers). They claim that nothing can warrant the existence of objects independent of our recognition, which many positivists admit without criticism (anti-realism). Therefore, we can only have a dialogue between subjects via text.

Some authors placed the simulation and game/role playing on the boundary of the empirical and constructive approaches (Galliers, 1991; Galliers and Land, 1987). The discussion covers the simulation used in ISR in general. If we focus the placement of the social simulation with agent-based approach, it is possible for us to reinforce their placement of the simulation on the boundary of dichotomy based on the following two reasons:

1. The subject–object dualism, which is the basis of the empirical approach, is not applicable to research objects of social simulation with the agent-based approach.
2. Simulations have roots in natural science, and the practical judgment of model validity from the standpoint of the empirical approach is still dominant.

We will examine these reasons in the next section in detail.

11.3 Social Simulation Research with an Agent-Based Approach

Autonomous agents in a society contain decision-making mechanisms. Decision-making mechanisms possess some recognition of the external world within them. This recognition of the world, which is formed inside a decision maker, is sometimes called an internal model. In an agent-based approach, a decision-making unit in some form with an internal model is called an agent, and modelers are supposed to give a computational model of its mechanism explicitly. Therefore, agents, which are targets of our modeling, include the activity of "recognizing" the external world within these agents. This cannot be avoided, even though a model is built to be simple enough under the KISS (keep it simple, stupid) principle, as long as we define an "agent" as an autonomous decision-making unit.

As long as we try to model heterogeneous agents as decision-making units, we cannot avoid the interaction between modelers and agents as our modeling targets. Not only our observations of agent decisions but also the agent decisions themselves are influenced by our investigation. Guba and Lincoln refer to this concept as "the interactive nature of the knower-known dyad" (Guba and Lincoln, 1989).

> Where the inquired into (we hesitate to call a human an "object") is a human, or a human characteristic, the existence of interconnectedness is inescapable, even if only at so primitive a level as the well-known phenomenon of reactivity. (Guba and Lincoln, 1989, p. 67)

If we accept that our observation and validation activity interact with the decision of the agents, then we can no longer base the validity of models on reality, which is independent of observation.

However, simulation, especially computer simulation, has its roots in natural science, the targets of which are physical phenomena. The existence of a real target system and its behavior are assumed, as are the traditional criteria, in which the validity of a model is judged based on the relation between the model and the real target. As demonstrated in the discussion between Windrum et al. (2007) and Moss (2008), this tendency can be seen in social science, especially in computational economics.

Nevertheless, as described above, since the social simulation with the agent-based approach is not based on the subject–object dualism, the agent-based approach must inevitably "approach" the epistemology of the constructivist. As mentioned in the previous section, the epistemology of the constructive approach focuses on the meaning of the social existence, which is constructed through intersubjective dialogue via text. Text is an expression of social existence in some form for constructivists. What are texts for agent-based modelers and researchers? We will suggest a working hypothesis as the answer to this question in order to proceed.

> Text for agent-based modelers is a simulation model.

Then, what is intersubjective dialogue via a model? We have to examine the role of simulation models in agent-based approaches in more detail in order to answer this question.

11.4 Role of Simulation Models in Agent-Based Approach

According to Richiardi, "the term 'validity' can be formally defined as the degree of homomorphism between one system and a second system that it purportedly represents" (Richiardi et al., 2006). The mathematical general systems theory (MGST) uses homomorphisms as the core concepts in discussing the relationship of systems formally (Mesarovic and Takahara, 1989). In his foundational work, Zeigler discusses the use of the homomorphism concept for validating simulation models (Zeigler et al., 2000). Unfortunately, agent-based modeling did not exist when he published the first edition of his book in 1976, and aspects that are specific to agent-based modeling were not explicitly discussed. We will discuss validity more rigorously from an MGST standpoint.

In a particular school of MGST called the logical approach or the model theoretical approach, a system, in general, can be defined as a mathematical structure (Takahashi and Takahara, 1995). A mathematical structure is a base set (in the sense of set theory) on which relations or functions (also in the sense of set theory) are defined.[1] A function can be defined as a relation of a special type. Therefore, more generally, a system can be defined as a base set on which relations are defined.

Let S and S' be two systems, such as $S = \langle A, \{R_i\}_{i \in I}\rangle$ and $S' = \langle A', \{R'_i\}_{i \in I}\rangle$, where A and A' denote base sets for respective systems. Here, A and A' are sets of elements that constitute two systems. Moreover, i denotes the index, which is the name of the structural property involved in both systems, and we assume that i is a member of the common index set I. For each $i \in I$, let R_i and R'_i be $\lambda(i)$-ary relations on A and A', respectively. In other words,

$$R_i \subset A^{\lambda(i)} \text{ and } R'_i \subset (A')^{\lambda(i)},$$

and $\lambda : I \to \mathbf{N}$ is the function defining the arity of the relation for each index (called the arity function). Here, $\mathbf{N}$ is the set of natural integers, and $A^{\lambda(i)}$ denotes the Cartesian product of A to the $\lambda(i)$-th power. We are considering the similarity between two systems. Therefore, we assume that these two systems are of the same type. In other words, S and S' have the same number of relations (the common index set I makes this hold), and relations with the corresponding index have the same arity (the fact that R_i and R'_i are both $\lambda(i)$-ary relations makes this hold).

A homomorphism from S into S' is a structure preserving function from S into S'. Strictly speaking, it is a mapping $h : A \to A'$ from A into A' that satisfies the following condition for all $i \in I$ and $x_1, x_2, \ldots, x_{\lambda(i)} \in A$:

$$(x_1, x_2, \ldots, x_{\lambda(i)}) \in R_i \Rightarrow (h(x_1), h(x_2), \ldots, h(x_{\lambda(i)})) \in R'_i.$$

In other words, elements that are in the relation R_i in the system S are also in the relation R'_i in the system S' when h maps these elements into S'. Henceforth, we write $h : S \to S'$ for a homomorphism $h : A \to A'$ from S into S'.

Now, we can give the formal definition of validity using the homomorphism concept as follows.

Definition 1 (Validity) Let S and S' be arbitrary systems with the same index set and arity function. Here, S' is said to be valid w.r.t. S when there is a homomorphism h from S into S'. □

Note that the homomorphism between S and S' is not necessarily unique. Also note that we are using homomorphism, not isomorphism. A homomorphism h can be a *many-to-one* mapping from S *into* S'. Therefore, there might be some sort of

[1] Although some authors (Enderton, 1972; Gallier, 1986; Sperschneider and Antoniou, 1991) assume multiple sets for one structure, we assume a single set here. The single set definition can be extended to a multiple set definition easily, and there is no essential difference between these definitions.

reduction from S into S'. Moreover, the image of h does not necessarily cover the entirety of A'.[2] Hence, there might be some redundancy in S'.

For example, let A be a set of objects located in a geographical region, and let A' be a set of objects on a map, such as roads, buildings, rivers, or hills. Let R and R' be a binary relation of "1 kilometer away in horizontal distance" and a binary relation of "10 centimeters away on the map," respectively. Moreover, let h be a function that maps objects in the region to objects described on the map. Then, h becomes a homomorphism whenever the map is described "correctly." Therefore, the map is said to be valid w.r.t. the region. Note that h maps two distinctive objects in A into identical objects in A'. For example, three-dimensional space is reduced to a two-dimensional expression on the map. Therefore, one room on the first floor and another room on the second floor of the same building in A can be mapped to the same object in map A'. There is additional information in map A' that does not have a correspondent in A, such as colors or state borders.

Models in the agent-based approach category are of course expressions of systems. Therefore, they can be handled within the framework of mathematical structures described above. Hereafter, we will express a model in the form $M = \langle A, \{R_i\}_{i \in I} \rangle$, where A contains potential time functions describing the attributes of agents, and R_i contains functions in the form of relations describing the dynamics of agents, for example.

We can give the formal definition of model validation concepts from two world views described above.

11.4.1 Empirical Validity

In the traditional empirical (positive) world view, the objective system $\overline{S} = \langle \overline{A}, \{\overline{R}_i\}_{i \in I} \rangle$, which is the target of modeling, is assumed, and the validity is judged on the relationship between a model and $\overline{S}$.

Definition 2 (Empirical Validity) Here, M is empirically valid if M is valid with respect to $\overline{S}$. □

---

[2] The image of $h : A \to A'$, sometimes denoted by $h(A)$, is the subset of A' defined as follows:

$$h(A) = \{h(x) | x \in A\}.$$

In other words, the image of h is the set of all $h(x)$ (the image of x under h) when x covers the entirety of domain A.

11.4.2 Constructive Validity

In the constructive world view, a subject has to judge validity based on the relationship between a model as text and the system that resides in each subjective view of a system. We think that both the worldview of the subject and its expression can be treated as systems. Each subjective worldview, called a subjective system, cannot be expressed explicitly, and other subjects cannot refer to this system directly. Hereinafter, we denote a subjective system in subject j as $S^j = \langle A^j, \{R_i^j\}_{i \in I} \rangle$. As described above, although S^j cannot be directly referenced by other subjects, j can.

Definition 3 (Constructive Validity) Assume a finite set of interpreters $J = \{1, 2, \ldots, n\}$ and a set of subjective models of those interpreters $\{S^1, S^2, \ldots, S^n\}$. Here, M is constructively valid with respect to the interpreter set J if M is valid with respect to S^j for each $j \in J$. □

This concept of validity is depicted in Fig. 11.1.

11.4.3 Intersubjective Dialogue via Models

Based on the definition of constructive validity, we can draw the picture of intersubjective dialogue via models. Assume there are two persons A and B and let their subjective systems be $S_A^{(1)} = \langle A^{(1)}, \{Q_i^{(1)}\}_{i \in I} \rangle$ and $S_B^{(1)} = \langle B^{(1)}, \{R_i^{(1)}\}_{i \in I} \rangle$, respectively. Although it is not obvious that we can put the identical index set I for both A and B, it is possible to assume that these persons have some elements of the structural aspects of society in common because they are about to discuss society

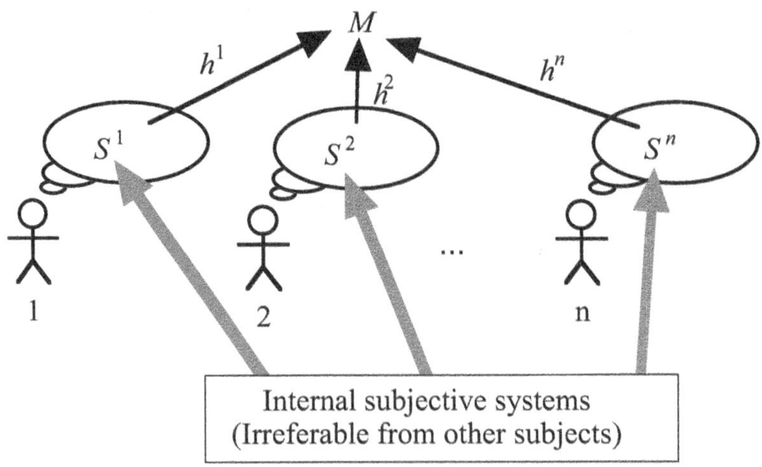

Fig. 11.1 Constructive validity

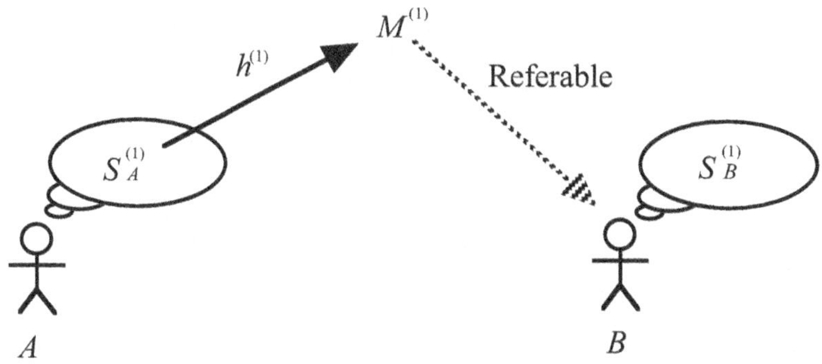

Fig. 11.2 Intersubjective dialog—step 1

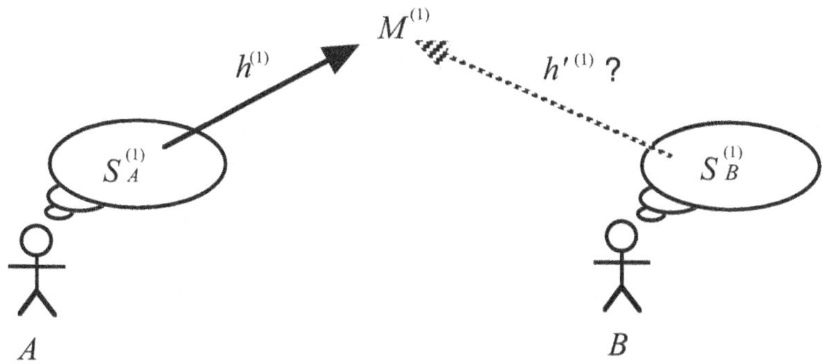

Fig. 11.3 Intersubjective dialog—step 2

over a simulation model. Therefore, we can adopt the intersection of these index sets for common I. The superscript (1) denotes that these systems are the first version of subjective systems for each subject. Hereinafter, we omit the index set I.

Assume that A expresses a valid model $M^{(1)} = \langle C^{(1)}, \{T_i^{(1)}\}\rangle$ w.r.t. his/her subjective system $S_A^{(1)}$. Here, $M^{(1)}$ is explicitly expressed as a text. Therefore, $M^{(1)}$ can be referenced by B (Fig. 11.2) and is assumed to be valid w.r.t. $S_A^{(1)}$. Hence, from the above definition of validity, there is a homomorphism $h^{(1)} : S_A^{(1)} \to M^{(1)}$. When B interprets the model $M^{(1)}$, B examines whether $M^{(1)}$ is valid w.r.t. his/her own subjective system $S_B^{(1)}$. In the case that there is a homomorphism, the homomorphism is not necessarily unique and there can be multiple homomorphisms in general. Therefore, B examines whether he/she could find an $h'^{(1)}$ of the homomorphisms (Fig. 11.3). If this is possible, then A and B successfully share a commensurable point via the model $M^{(1)}$ (Fig. 11.4). If B cannot find an $h'^{(1)}$ of

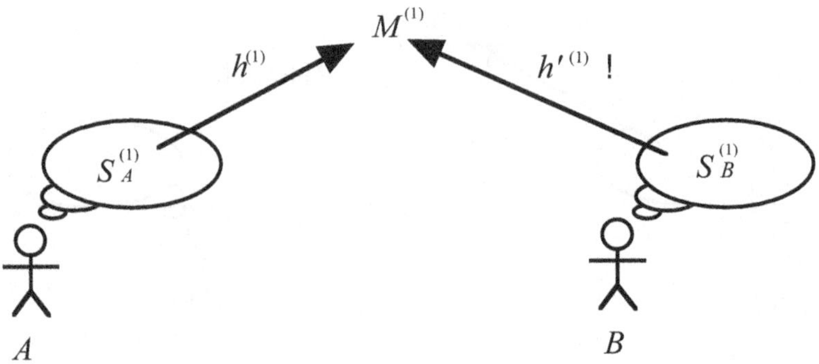

Fig. 11.4 Intersubjective dialog—step 3

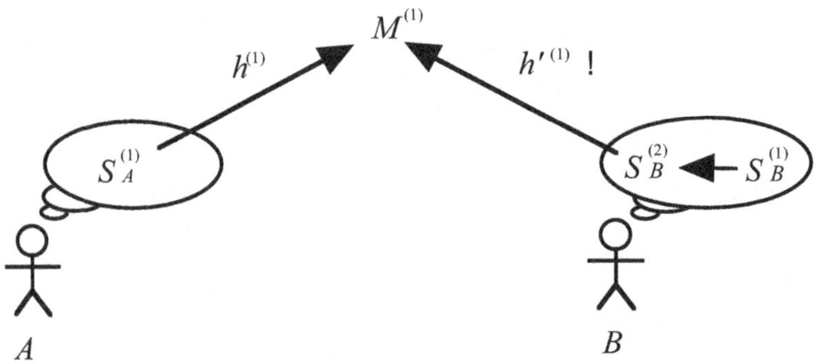

Fig. 11.5 Intersubjective dialog—step 4

the homomorphisms, then B judges that $M^{(1)}$ is not valid and one of the following three options can be selected:

First, B revises his/her subjective system into $S_B^{(2)}$ so that he/she could make a homomorphism into the model $M^{(1)}$. If he/she can find a new homomorphism $h'^{(1)} : S_B^{(2)} \to M^{(1)}$ with this revision, then the dialogue of two subjects is successful (Fig. 11.5). In this case, the model $M^{(1)}$ might be examined further by other interpreters.

However, if B fails in finding a homomorphism even after revising the model into $S_B^{(2)}$, as the second option, B can express a valid model $M^{(2)} = \langle C^{(2)}, \{T_i^{(2)}\}\rangle$ w.r.t. his/her newly revised subjective system $S_B^{(2)}$ (Fig. 11.6). Here, $M^{(2)}$ is assumed to be a valid model w.r.t. $S_B^{(2)}$. Therefore, there exists a homomorphism $h^{(2)} : S_B^{(2)} \to M^{(2)}$, as was the case of the valid $M^{(1)}$ of A above. In addition, as a next step, the new second version model $M^{(2)}$ is interpreted by A. If A and B succeed in sharing a valid model w.r.t. both of their subjective systems by repeating this process, then A

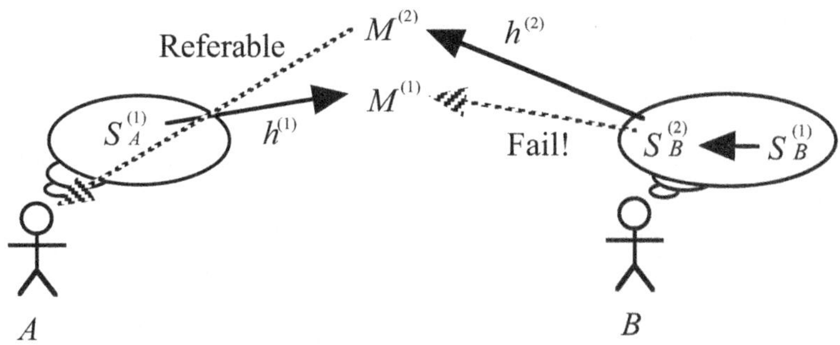

Fig. 11.6 Intersubjective dialog—step 5

and B can reach a common understanding of the society through the intersubjective dialogue between A and B via models as text.

As the third option, B can reject $M^{(1)}$ by judging the model to be invalid when it is expressed by A. In this case, B can express the second version of model $M^{(2)}$, which is valid w.r.t. the first version of his/her subjective system $S_B^{(1)}$ without revision, and $M^{(2)}$ is to be interpreted and judged for validity by A as the next step. Of course, there is a possibility that both A and B reject the model of the other by judging the models to be invalid, and their dialogue can break down (incommensurable situation).

We might think that similar analysis can be performed with a positivist world view if one accepts that indicators (e.g., measurements) are theory-laden. Within a positivist world view, however, the relation between the objective system (denoted as $\overline{S}$ in the previous section) and a model (M) is discussed, rather than the relation between the subjective systems (S^j) and a model (M). The theory-ladenness makes multiple correspondences (h) between the objective system and a model possible. Researchers with different theories can argue for valid correspondence (homomorphism) between $\overline{S}$ and M.

11.4.4 Mixed Validity

It is difficult for a practical agent-based model to claim validity under the extreme standpoint of each of the two worldviews described above. We use these two validity concepts by "mixing" them depending on the objectives of simulation.

Many social simulation researchers appear to state the validity of a model for empirical validity on some parts and for constructive validity on other parts. Formally, we partition the set I, which is the set of names for structural properties involved in a model, into two sets that denote on which of the two types of validity each property is based.

Definition 4 (Structure Partitioning) Given a model $M = \langle A, \{R_i\}_{i\in I}\rangle$, assume that I is partitioned into two sets, I_E and I_C. Hence,

$$I = I_E \cup I_C, \quad I_E \cap I_C = \emptyset.$$

I_E, I_C are index sets representing structural properties with empirical validity and constructive validity, respectively. In the general definition of the partition, I_E and I_C are assumed to be non-empty, although we permit either of I_E or I_C to be empty. A pair (I_E, I_C) of these two sets is called a structural partition of the model. □

Given a model $M = \langle A, \{R_i\}_{i\in I}\rangle$ and a structural partition (I_E, I_C) of M, we can consider the following two submodels, the relations of which are restricted to I_E or I_C, respectively:

$$M_E = \langle A, \{R_i\}_{i\in I_E}\rangle,$$

$$M_C = \langle A, \{R_i\}_{i\in I_C}\rangle.$$

This situation is depicted in Fig. 11.7 with an example of some elements in I.

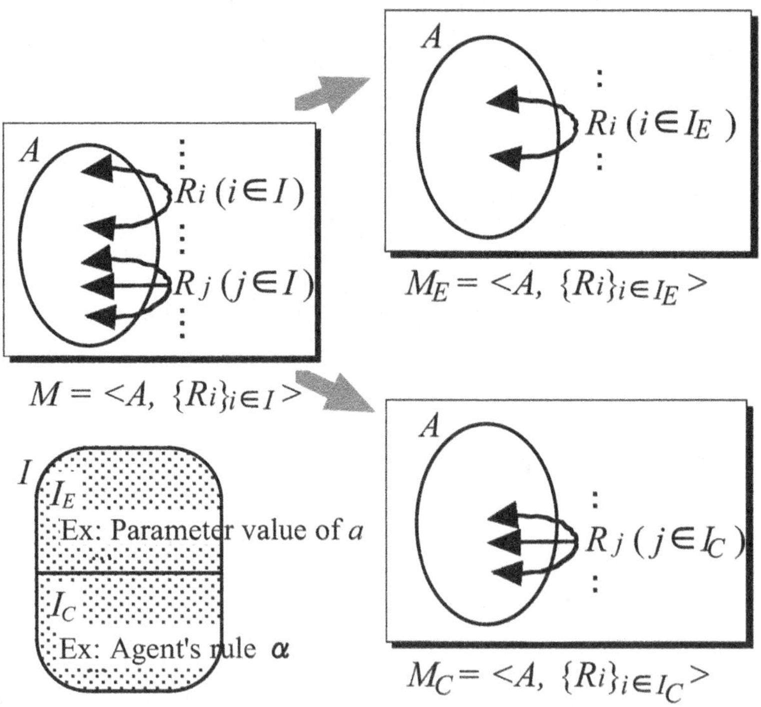

Fig. 11.7 Two induced models with structural partitioning

Definition 5 (Mixed Validity) Let a model $M = \langle A, \{R_i\}_{i \in I} \rangle$ and a structural partition (I_E, I_C) of M be given. Suppose the real system $\overline{S} = \langle \overline{A}, \{\overline{R}_i\}_{i \in I_E} \rangle$, which is a target of modeling, and suppose a finite set of interpreter $J = \{1, 2, \ldots, n\}$ and their subjective models $\{S^1, S^2, \ldots, S^n\}$, where for each $j \in J$,

$$S^j = \langle A^j, \{R_i^j\}_{i \in I_C} \rangle.$$

The model M is said to be empirically and constructively valid with respect to interpreter set J when the following two conditions hold:

1. The model M_E is valid with respect to $\overline{S}$, i.e., for all properties in I_E, M_E is empirically valid.
2. For each $j \in J$, M_C is valid with respect to S^j, i.e., for all properties in I_C, M_C is constructively valid with respect to interpreter set J. □

This concept is depicted in Fig. 11.8.

The discussion of validity gives more weight to either empirical validity or constructive validity, depending on whether I_E or I_C in the structure partition (I_E, I_C) of M has the greater number of elements. The model validity discussion can be placed somewhere on the spectrum depending on how the partition is made, from extreme empirical validity (in case that $I_E = I$ and $I_C = \emptyset$) to extreme constructive validity (in case that $I_E = \emptyset$ and $I_C = I$). This idea is depicted in Fig. 11.9.

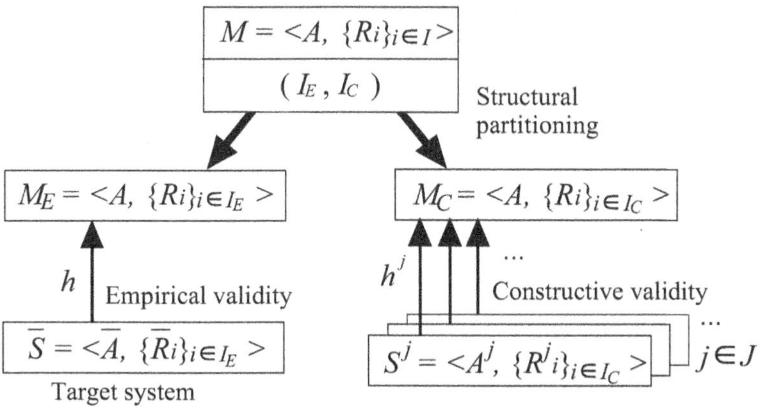

Fig. 11.8 Mixed validity

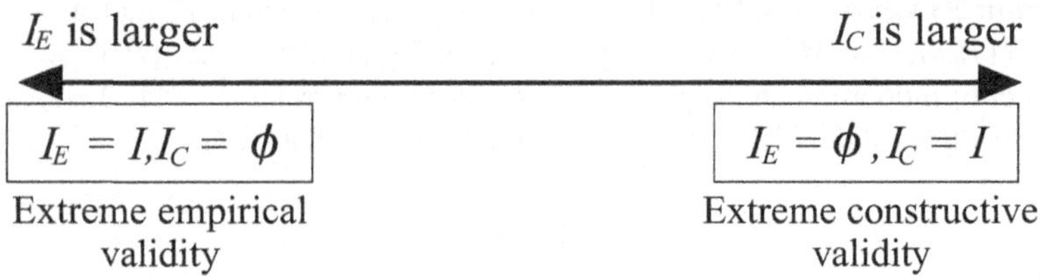

Fig. 11.9 Empirical/constructive validity spectrum

11.5 Conclusion

We need a novel perspective for the validation of agent-based approach models. The validation of models can be performed by mixing methods based on empirical and constructive standpoints. From the empirical standpoint, validity is defined as the relationship between a model as a system and an existing system in reality. From the constructive standpoint, validity is evaluated as the intersubjective dialogue between an author and a reader via models as text.

As discussed in Sect. 11.3, we have to rely on intersubjective dialogue via models to some extent for their validation in the agent-based approach. For this purpose, we have to share simulation models for intersubjective discussions. This means that it is an essential requirement for researchers to disclose their models to audiences (interpreters) as text. The most desirable method by which to do this is to share the executable code itself with audiences who have the same simulation platform as the authors. Using the Internet can be the most feasible method, as suggested by Gilbert and Troitzsch.

> "One solution is to publish the code itself on the Internet. A more radical solution is to publish in one of the increasing number of electronic journals" ("Simulation for the Social Scientist", by Gilbert and Troitzsch, 2005, p. 26).

Practically, not all social simulation researchers share a single common simulation platform, which makes sharing executable codes among researchers more difficult. As a compromise, models should be presented with formalism so that models can be replicated on other modeling platforms.

References

Arnold, E. (2019). Chapter 8: Validation of Computer Simulations from a Kuhnian Perspective. In C. Beisbart & N. J. Saam (Ed.), *Computer simulation validation* (pp. 203–224). Springer-Verlag.

Boero, R., & Squazzoni, F. (2005). Does empirical embeddedness matter? Methodological issues on agent-based models for analytical social science. *Journal of Artificial Societies and Social Simulation*, *8*(4), 6.

Enderton, H. B. (1972). *Mathematical introduction to logic*. Academic Press.

Fagiolo, G., Moneta, A., & Windrum, P. (2007). A critical guide to empirical validation of agent-based models in economics: Methodologies, procedures, and open problems. *Computational Economics*, *30*, 195–226.

Farhoomand, A. F. (1987). Scientific progress of management information systems. *Data Base*, *19*(2), 48–56.

Gallier, J. H. (1986). *Logic for computer science: Foundations of automatic theorem proving*. Harper & Row Publishers.

Galliers, R. D. (1991). Choosing appropriate information systems research approaches. In H.-E. Nissen et al. (Eds.), *Information systems research* (pp. 327–345). North-Holland.

Galliers, R. D., & Land, F. F. (1987). Choosing appropriate information systems research methodologies. *Communications of the ACM*, *30*(11), 900–902.

Gilbert, N., & Troitzsch, K. (2005). *Simulation for the social scientist* (2nd ed.). Open University Press.

Grimm, V., et al. (2005). Pattern-oriented modeling of agent-based complex systems: Lessons from ecology. *Science*, *310*(11), 987–991.

Guba, E. G., & Lincoln, Y. S. (1989). *Forth generation evaluation*. Sage Publications.

Hamilton, S., & Ives, B. (1982). MIS research strategies. *Information & Management*, *5*, 339–347.

Hirschheim, R., et al. (1995). *Information systems development and data modeling*. Cambridge University Press.

Kueppers, G., & Lenhard, J. (2005). Validation of simulation: Patterns in the social and natural sciences. *Journal of Artificial Societies and Social Simulation*, *8*(4), 3.

Law, A. M. (2007). *Simulation modeling and analysis* (4th ed.). McGraw-Hill.

Lee, A. S. (1991). Architecture as a reference discipline for MIS. In H.-E. Nissen, et al. (Eds.), *Information systems research* (pp. 573–592). North-Holland.

Mesarovic, M. D., & Takahara, Y. (1989). *Abstract systems theory*. Lecture notes in control and information sciences. Springer-Verlag.

Mingers, J., & Stowell, F. (Eds.). (1997). *Information systems: An emerging discipline?*. McGraw-Hill.

Morone, P., & Taylor, R. (2010). *Knowledge Diffusion and Innovation*. Modelling complex entrepreneurial behaviours. Edward Elgar Pub. Ltd.

Moss, S. (2008). Alternative approaches to the empirical validation of agent-based models. *Journal of Artificial Societies and Social Simulation*, *11*(1), 5.

Naylor, T. H., & Finger, J. M. (1967). Verification of computer simulation model. *Management Science*, *14*(2), B92–B101.

Oreskes, N., Shrader-Frechette, K., & Belitz, K. (1994). Verification, validation, and confirmation of numerical models in the earth sciences. *Science*, *263*(4), 641–648.

Railsback, S. F., & Grimm, V. (2019). *Agent-based and individual-based modeling: A practical introduction* (2nd ed.). Princeton University Press.

Richiardi, M., et al. (2006). A common protocol for agent-based social simulation. *Journal of Artificial Societies and Social Simulation*, *9*(1), 15.

Sargent, R. G. (2011). Verification and validation of simulation models. In S. Jain, et al. (Eds.), *Proceedings of the 2011 Winter Simulation Conference* (pp. 183–198). IEEE.

Shiba, N. (2019). The validity of agent-based models (in Japanese). In *The 10th Conference of Transdisciplinary Federation of Science and Technology, F-4*.

Sperschneider, V., & Antoniou, G. (1991). *Logic: A foundation for computer science*. Addison-Wesley Pub. Co.

Sun, J., & Tesfatsion, L. (2007). Dynamic testing of wholesale power market designs: An open-source agent-based framework. *Computational Economics*, *30*, 291–327.

Takahara, Y., & Liu, Y. (2006). *Foundations and applications of MIS: A model theory approach*. Springer.

Takahashi, S., & Takahara, Y. (1995). *Logical approach to systems theory*. Lecture notes in control and information science. Springer.

Vogel, D. R., & Wetherbe, J. C. (1984). MIS research: A profile of leading journal and universities. *Data Base*, *16*(3), 3–14.

Werker, C., & Brenner, T. (2014). *Empirical calibration of simulation models*, Papers on Economic and Evolution No. 0410, Max Planck Institute for Research into Economic Systems.

Windrum, P., Fagiolo, G., & Moneta, A. (2007). Empirical validation of agent-based models: Alternatives and prospects. *Journal of Artificial Societies and Social Simulation, 10*(2), 8.

Zeigler, B. P., Praehofer, H., & Kim, T. G. (2000). *Theory of modeling and simulation* (2nd ed.). Academic Press.

The manufacturer's authorised representative in the EU is Springer Nature Customer Service Centre GmbH, Europaplatz 3, 69115 Heidelberg, Germany. If you have any concerns regarding our products, please contact ProductSafety@springernature.com

Printed and bound by CPI Group (UK) Ltd, Croydon, CR0 4YY
30/09/2026
02225329-0006